Ringing in the Common Love of Good
The United Farmers of Ontario, 1914–1926

Founded in 1914, the United Farmers of Ontario (UFO) became a significant force in the province, winning the most seats in the 1919 provincial election and forming a governing coalition with the Independent Labour Party. The UFO and its companion organizations, the United Farmers Cooperative Company (UFCC) and the United Farm Women of Ontario (UFWO), flourished, achieving much of its success by challenging those who controlled the economic, political, and social structures in Ontario and advancing an alternative vision of democracy that sought to maximize citizen participation in the decision-making process.

By the mid-1920s the UFO had gone into a period of decline from which it never recovered. The promise of equality hoped for by UFWO members never materialized and the UFCC, once a key component in the development of an alternative vision, began to focus more on profits than on politics.

In *Ringing in the Common Love of Good* Kerry Badgley explores both the rise and the fall of the UFO, focusing on the Ontario counties of Lambton, Simcoe, and Lanark. He challenges the liberal-capitalist interpretation that the movement was nothing more than a group of impatient Liberals, as well as the Marxist view that the UFO consisted of self-interested independent commodity producers. Badgley argues that as the UFO broke free from hegemonic forces it developed alternative economic, political, and social visions, but that it was these same forces, combined with internal struggles and a conservative leadership, that ultimately resulted in the decline of the movement as a vehicle for democratic change in Ontario.

KERRY BADGLEY is an archivist at the National Archives of Canada.

Ringing in the Common Love of Good

The United Farmers of Ontario, 1914–1926

KERRY A. BADGLEY

McGill-Queen's University Press
Montreal & Kingston · London · Ithaca

ISBN 0-7735-1895-9

Legal deposit first quarter 2000
Bibliothèque nationale du Québec

Printed in Canada on acid-free paper

This book has been published with the help of a grant from the Humanities and Social Sciences Federation of Canada, using funds provided by the Social Sciences and Humanities Research Council of Canada.

McGill-Queen's University Press acknowledges the financial support of the Government of Canada through the Book Publishing Industry Development Program for its activities. We also acknowledge the support of the Canada Council for the Arts for our publishing program.

Canadian Cataloguing in Publication Data

Badgley, Kerry, 1962–
Ringing in the common love of good : the United Farmers of Ontario in Lambton, Simcoe and Lanark Counties, 1914–1926
Includes bibliographical references and index.
ISBN 0-7735-1895-9
1. United Farmers of Ontario – History. 2. Ontario – Politics and government – 1905–1919. I. Title.
HD1486.C3U66 2000 324.2713'02 C99-901046-8

Typeset in Palatino 10/12
by Caractéra inc., Quebec City

Contents

Tables and Figures

TABLES

FIGURES

Abbreviations

AG	*Almonte Gazette*
AO	Archives of Ontario
BE	*Barrie Examiner*
BNA	*Barrie Northern Advance*
CB	*Collingwood Bulletin*
CE	*Collingwood Enterprise*
CPH	*Carleton Place Herald*
CSB	*Collingwood Saturday Bulletin*
CSN	*Collingwood Saturday News*
FA	Farmers' Association
FFP	*Forest Free Press*
FI	Farmers' Institute
FS	*Forest Standard*
ILP	Independent Labor Party
LCA	Lambton County Archives
LE	*Lanark Era*
MFP	*Midland Free Press*
NA	*Northern Advance*

OAC	Ontario Agricultural College
OP	*Orillia Packet*
OPT	*Orillia Packet and Times*
OSA	*Oil Springs Advance*
OSN	*Orillia Saturday News*
OT	*Orillia Times*
OWT	*Orillia Weekly Times*
PAT	*Petrolia Advertiser-Topic*
PC	*Perth Courier*
PE	*Perth Expositor*
POI	Patrons of Industry
PT	*Petrolia Topic*
SCO	*Sarnia Canadian Observer*
SFRN	*Smiths Falls Record-News*
SFRR	*Smiths Falls Rideau Record*
SUN	*Farmers's Sun*
UFO	United Farmers of Ontario
UFWO	United Farm Women of Ontario
WI	Women's Institute

Acknowledgments

It is impossible to thank everyone who helped me complete this book. There were, however, some whose efforts went well beyond the call of duty; and they must be mentioned.

Professor J.K. Johnson, who supervised the thesis upon which this book is based, was of immense assistance. He carefully read and commented on earlier versions of this study, forcing me to rethink or clarify some of my ideas. Elizabeth Price read drafts of each chapter and made many constructive suggestions as to how they could be improved. She also listened to my ramblings on the UFO and still remained my friend. Her words of encouragement are appreciated more than she will ever know.

Staff at the Archives of Ontario were very helpful, as were those at the Simcoe County Archives, Lambton County Archives, and University of Guelph Library. I also wish to thank my editor at McGill-Queen's, Roger Martin, who guided me through what otherwise would have been a daunting publication process and Claire Gigantes, my copy editor, whose suggestions improved my prose.

Several of other people also assisted me (knowingly and otherwise) in completing this book, including N.E.S. Griffiths, Duncan McDowall, D.A. Muise, Carman Bickerton, Robert McIntosh, Irene M. Spry, D.H. Akenson, Bill Ramp, Brett Fairbairn, D.C. Savage, Nancy Kiefer, Susan Paré, Simon Snow, Shawn Cafferky, Charlene Porsild, John Franklin, Dan German, Larry Richer, Tony Bonacci, David Shanahan, Mary McLeod, Pat Milliken, Jim Kenny, Grant and Barbara Tayler, Alex Sim, and D.S. Macmillan. Thanks are also extended to my brother, Aaron, for his encouragement. Needless to say, I am solely responsible for any errors and omissions.

Finally, three people who deserve special mention. My wife, Susan, endured more than anyone should have to while I worked on this study. Through it all she gave me encouragement and love, and showed me why it was so important that I finish it. Our daughter Sarah arrived just as the project was nearing completion and provided yet another reason for finishing it. Finally, none of this would have been possible without my mother, Marion, who did not live to see this book. A single working parent through much of my childhood, she managed to find the time to inculcate me with a love for learning and to show me time and again why so-called "common" people deserve to be treated historically. It is to my mother that this book is dedicated.

"Ring out the narrowing lust of gold;
ring in the common love of good"

Headline in the pro-UFO *Almonte Gazette* on the eve of the 1921 federal election, 2 December 1921. Quoting a hymn by Alfred, Lord Tennyson, 1850

1 Introduction

> To ask questions about the nature and behavior of one's own society is often difficult and unpleasant: difficult because the answers are generally concealed and unpleasant because the answers are often not only ugly … but also painful … In contrast, the easy way is to succumb to the demands of the powerful, to avoid searching questions, and to accept the doctrine that is hammered home incessantly by the propaganda system.
>
> Noam Chomsky[1]

What follows is an account of how a large group of people in Ontario came to see things differently, and of how they then lost their alternative vision. Known as the United Farmers of Ontario (UFO), this group formulated a critique of society that had potentially profound implications. They questioned economic system; raised doubts about the effectiveness of existing political structures; expressed uncertainty regarding the treatment of women and ambivalence with respect to war, militarism, and patriotism. They were suspicious of those who disseminated information in society, particularly journalists and politicians. In sum, UFO members rejected many of the prevailing societal values. They were dissidents.

Although they initially acted with spontaneity, fervour, and conviction, members of the UFO could not sustain their societal critique, and their ability to see things differently evaporated in fairly short order. It is the process of their acquiring and then losing the capacity to posit a vision of society that is addressed here.

In tracing the emergence and virtual disappearance of the UFO's vision, I make the following contentions. First, the movement represented a potentially significant challenge to the established order, a challenge that was spontaneous, highly creative, and democratic in its outlook. Idealism was a vital component of the UFO, and it distinguished its rank-and-file members from many of their contemporaries. "Potentially" should be stressed, however, because from its inception the movement faced opposition from mainstream society, whose co-optive tactics weakened the UFO's actions. Second, at the local level UFO members did not always agree with the central leadership. There were moments, however, when local club members relied on their leaders for advice and direction. The effect of such reliance was profound because many influential figures in the UFO, their rhetoric notwithstanding, adopted "Liberal" rather than radical positions. Third, despite some scholars' claims to the contrary, sincere and creative efforts were made by UFO members to unite with labour's contemporary political movement, the Independent Labor Party (ILP). Fourth, despite their ability to break free from certain fundamental societal assumptions, UFO members were unable to extricate themselves completely from others, as the experience of women in the movement shows. Finally, one of the more important factors accounting for the decline of the movement was the failure of its cooperative wing to sustain an alternative to prevailing economic practice. Indeed, the UFO's cooperative activities provide insights on how hegemony can redirect the response of a mass democratic movement.

THREE COUNTIES

In order to test these contentions, I look at three counties – Lambton (in southwestern Ontario), Simcoe (in the central portion of the province), and Lanark (in eastern Ontario) – that represent three distinct regions within "Old Ontario." By studying these counties, it is possible to determine whether or not there were any regional variations in the UFO. In Lambton and Simcoe, local records pertaining to the UFO have survived. All three counties were reasonably well served by local newspapers, and from this source it has been possible to reconstruct, at least in outline, the major UFO clubs, thus providing sufficient information on which to base an analysis.[2]

THE UFO

The UFO was formed in 1914 through the efforts of W.C. Good, J.J. Morrison, E.C. Drury, and Col J.Z. Fraser, four prominent Ontario agrarians. Their intention was to create an organization – a pressure group, not a party – that would serve the educational, economic, and political needs of farmers.

The UFO was part of the tradition of agrarian activism that was reflected in the Patrons of Husbandry (better known as the Grange), a US-based secret society that entered the province in 1874. Some farmers, dissatisfied with the Grange's refusal to enter politics, later became members of another US-based secret society, the Patrons of

Industry (POI), which entered Ontario in 1889. This group elected seventeen members to the Ontario Legislature in the 1894 election and came close to electing members in twenty other ridings.[3] Soon after the election, however, the POI imploded and was a non-entity by 1900. A farm lobby group, the Farmers' Association (FA) was established in Ontario in 1902 and enjoyed an initial measure of success.

By 1914, however, Ontario farmers' organizations were in a state of disarray; the FA had effectively ceased to exist, and the Grange was a shadow of its former self. By 1914 the largest agrarian body in Ontario was the Farmers' Institute (FI), a provincially sponsored educational organization. In fact, several Farmers' Clubs, which were part of the FI, were "stolen" by J.J. Morrison and others when the UFO was formed; they convinced these clubs to drop their government ties and to join instead with the UFO.[4] Initially, response to the organization was lukewarm, but with the arrival of the First World War and the convergence of other political, social, and economic forces, support for the UFO grew exponentially. When it entered the arena of electoral politics in 1918, the movement assumed a new form. Farmers in the riding of Manitoulin, acting without advice or guidance from the central organization, nominated one of their own to contest a provincial by-election and won the seat. Another by-election victory soon followed in the riding of Ontario North, adding to the farmers' self-confidence and making them more willing to challenge some of the fundamental assumptions of their society.

At the same time, the UFO's sister organization, the United Farmers' Cooperative Company (UFCC), entered into a number of cooperative ventures. A central buying-and-selling agency enabled local clubs to purchase farm supplies and market their produce at lower costs than through traditional "middlemen." Later, a short-lived chain of grocery stores made an appearance.[5]

The movement's popularity culminated in the 1919 provincial election (in which the UFO won a plurality of seats and formed a coalition government with the ILP) and with the performance of the Progressive party (the federal wing of the agrarian movement) in the 1921 federal election (in which it elected twenty-four Ontario MPs out of a caucus of sixty-five).[6] By that time, the UFO could boast of having some 60,000 members, including members of the United Farm Women of Ontario (UFWO) and the United Farm Young People of Ontario (UFYPO).[7]

Despite its success, the UFO quite rapidly went into decline. With the exception of support for its cooperative activity, membership and interest had fallen off dramatically by the mid-1920s. Politically, the movement suffered a decisive loss in the 1923 provincial election, and it was unable to replicate its 1921 federal performance in the 1925 and 1926 electoral contests. The UFO continued to operate, in a reduced form, until it disbanded in 1943.

Aside from a few contemporary accounts of the UFO,[8] autobiographies of two important figures in the movement,[9] and a few (mainly unpublished) biographies of the organization's leaders,[10] relatively little has been written on the United Farmers of Ontario. With a few exceptions, what does exist dates primarily from the 1960s and 1970s. A survey of this literature demonstrates that, although certain themes have been explored in great detail, other equally important aspects of the movement have been all but ignored.

One theme that has received some attention is the rise and fall of the UFO. Some scholars point to the government's cancellation of promised conscription exemptions for rural workers as a primary impetus for the growth of the UFO.[11] To conscription issues W.R. Young adds the changing social character of the province from a rural to an urban society.[12]

As for the decline of the UFO, Wayne C. Brown argues that it had more to do with the "broadening out" controversy than with other factors. For Terry Crowley it was linked to the prosperity the farmers enjoyed as the 1920s progressed. Foster Griezic believes that the movement lost support because of "severe post-war agricultural depression from 1920 to 1925."[13]

All of the elements that historians have cited – conscription, social change, economic conditions, and so on – contributed to the rise and fall of the UFO, but they were augmented by other factors. The growth of the movement was related to a sense that the promise of democracy was unfulfilled in the province, despite the claims of the traditional parties. In the face of the slaughter in Europe, profound scepticism was expressed at the local level about the efficacy of Ontario's political structures. More importantly, these attitudes began to feed into other aspects of farmers' lives. When, for example, the federal government of Prime Minister Robert Borden revoked conscription exemptions for agricultural labourers, several spontaneous and militant meetings took place along concession lines. At these meetings farmers, although they were labelled traitors by the press, began to realize that if they banded together there might be a chance of effecting real change in the polity.

The decline of the movement can be attributed, at least in part, to the dissonance created by the central UFO and by agrarian politicians. Members who had supported UFO and Progressive candidates became convinced that things were not going to change substantially as a result of electoral victories, and they began to believe that there was no advantage to be had in supporting these politicians.

One also sees in the experience of the UFO the power of hegemony in political culture. Farmers broke free from hegemonic forces, but only momentarily. They could not sustain the effort in the face of the oppositional elements that acted vehemently to counter their dissent. Equally important, forces from within the movement, especially at the central level, actually aided the opposition against it. Altogether the weight of hegemony became too much for local members to bear, and many simply stopped trying to present an alternative vision.

The most problematic aspect of existing studies of the UFO is that they all focus on the leaders and pay only passing notice to local members.[14] This is a curious treatment for such an avowedly decentralist movement. "Constituency autonomy" was one of the guiding principles of the UFO, and it was taken seriously. Ultimately, it was the decline in members and support at the local level that led to the UFO's downfall. Even if they occasionally deferred to the central organization, the local clubs were very important, and an examination of when and why they deferred helps explain the decline of the movement. Hence, I hope to begin the process of examining the UFO at a level that has hitherto been largely ignored.

THE UFO AND ONTARIO HISTORIOGRAPHY

In order to explain how the UFO could be treated in so consistently unsatisfactory a manner, some mention must be made of the historiographic paradigms that pervade the study of Ontario.

The most prevalent framework used for explaining the province's history is one that points to Ontario political culture as being inherently conservative.[15] This conservatism, supposedly developed in the pre-Confederation era, emphasizes rule by élites and an aversion to democratic practice. By the early twentieth century, according to this paradigm, these attitudes had filtered down from the élites to the extent that they were shared by the average citizen.[16] While desiring change, Ontarians apparently want it to be gradual and pragmatic, with minimal disruption to the *status quo*.[17] Those in power who deviate from this path are punished with defeat at election time.

This interpretation is pervasive. For instance, those who study the immediate post-Confederation period write of Premier Oliver Mowat's pragmatism and conservatism. The early part of the twentieth century belongs to the progressive but pragmatic Premier James

Whitney.[18] Such traits are seen in all provincial governments until 1919, when the first deviation occurred. Aside from the Drury and Hepburn interludes, the tradition continues, or so the argument goes, to the present day.

Thus, the defeat of the UFO came naturally enough because of the province's conservative disposition and because of the natural inclination to revert to the "brokerage" model of politics. Nowhere is it mentioned that the UFO/ILP coalition may have been defeated because it did not go far enough in its agenda. What occurred at the local level at the time, however, suggests that this may have been the case.[19]

Clearly, there are problems with the conservative paradigm. First, in order to adhere to it one must necessarily ignore the instances in Ontario's past when significant challenges to élite dominance have occurred. To do so is no small feat, for it means downplaying or ignoring the Rebellion of 1837, the Grange, Knights of Labor, Patrons of Industry, UFO, Cooperative Commonwealth Federation (CCF), and the Communist party. In fact, one way to diminish the importance of these movements is to treat each one as a discrete occurrence, largely unconnected to any other.

A more nuanced approach to Ontario's political culture is provided by S.J.R. Noel's study of "clientelism," the pattern of "patron-client relationships that is woven into the total fabric of the community, and whose political effectiveness and durability are all the greater precisely because it is *not* exclusively political." In other words, the patron-client relationship is assumed to be a normal part of the political process, "because it was a normal part of practically everything else."[20] These relationships become pervasive to the point where they are not questioned; they are the normal state of affairs.

Noel concentrates primarily on the nineteenth century, ending his study with the end of the Mowat era in 1896. Despite the pervasiveness of clientelism, Noel points to a moment when a large number of people (POI supporters) rejected the system as it was and proposed instead a radically different one. This opposition was perceived as a serious threat to the existing order, and Noel describes Mowat's enormous effort to diffuse and eventually defeat it. He argues that the POI did not emerge from but rather was a challenge to clientelism, and he admits that his model cannot explain the movement's rise or its internal dynamics. However, Noel contends that the model does provide a good explanation for its rapid demise.[21]

A number of intriguing implications arise from Noel's work. First, he acknowledges (if only implicitly) that it took a great effort for POI members to free themselves from the prevailing political culture. In attempting to make this break, they were faced with a largely hostile press that simply could not entertain an alternative to either the Liberals or the Conservatives. Second, Noel points out that to reject clientelism was to reject the rewards that could be derived from entering into patron-client relationships. Noel and others consider the POI to have represented a significant challenge to the existing political structure. The movement's performance in the 1894 election stands as testimony to its strength. It is indeed curious, then, that so few observers consider the UFO *victory* in 1919 to be evidence of a similar challenge.

Equally problematic is the more recent historiographic model that claims that there are no coherent threads in Ontario's past. Reviewing the major historical works regarding the province that were written in the 1980s, David Gagan contends that they have "no unifying meaning or purpose beyond geographical reference." He argues that the studies help to describe "a unique society whose evolving characteristics were forged by processes that were in themselves not peculiar to Ontario. But in the context created by the interplay of time, place, people and circumstances they produced a distinctive society ... that has no precise antecedent and was not subsequently replicated in another time and another place as the nation expanded."[22] If Gagan's interpretative framework is the norm for present and future scholarly work, one cannot help but be discouraged on two counts. First, although there were moments in the province's past when responses to external forces were unique, this paradigm does not allow for those times when the response was *similar* to that of other regions or of other periods. This has more than passing relevance to students of populism, who are constantly told that Prairie populists were much more radical in their aims than their Ontario counterparts. Second, although there may be elements in the responses to external forces that have no discernible antecedents, one should not overlook the possibility that these responses were informed by past experience.

These are important issues, particularly since I argue here that no meaning can be derived from the past unless linkages are explored and continuity tested as the basis for analysis. It could be argued, for instance, that instead of being "crypto-Liberals" as they are labelled in some accounts, UFO members more closely resembled Prairie populists in their formulation of a creative and democratic response to the societal problems they identified. If this observation is taken to its logical conclusion, there is a connection to the historiographic paradigm that refers to Ontario's conservative political culture. Assuming that there was some continuity between UFO members and the earlier agrarian movements, and that these groups had links, however tenuous, to urban movements, a framework emerges that posits the notion that Ontario might not be the inherently conservative province it is so frequently described as being.

How, finally, have general accounts of Ontario's past treated the UFO? In some instances, agrarian protest in general and the UFO movement in particular have been treated with condescension, and in others they have been dismissed. There is, however, another common treatment – these subjects have been routinely addressed (and distorted) within the context of the supposed conservative nature of the province. Robert Bothwell, in his history of Ontario, argues that the UFO can be lumped together with the Liberal and Conservative parties in the period 1919–39 because it did not depart from orthodox political behaviour. Later, though, he notes that the UFO/ILP government of E.C. Drury provided "an object lesson in what can go wrong with a militant, one-issue reform party when it unexpectedly achieves power."[23] Bothwell's views resonate with others, but the most prevalent treatment of the UFO still remains to be mentioned: more often than not, farmer movements, and agriculture in general, are ignored.[24]

RURAL HISTORY

The lack of attention given to the UFO can be attributed in part to the present state of rural history. In a recent review of the field, R.W. Sandwell points out that several novel approaches have yielded work that challenges many commonly held assumptions. Rural historians are uncovering evidence that nineteenth-century farmers subordinated the accumulation of capital to a "pre-capitalist" social formation that gave priority to the independent family household. The implications of such findings are intriguing. If it can be shown that there was more to farming than mere money making, and if it was the case that this mentality carried over into the twentieth century, then the members of the UFO could be taken to be after something more than maximization of profits.[25]

Sandwell contends that what is needed is a better appreciation of the family farm as an "economic, political, and cultural institution." She argues that historians have traditionally paid lip service to this institution but have given only "scant attention to either the form of labour or the type of society characterized by non-wage workers organized on the basis of kinship within a capitalist system." More work needs to be done, Sandwell stresses, to enhance our understanding of the complex forces that acted on the family farm unit and of how they were "tied to the political and economic dynamics of the larger society." In fact, Sandwell argues that if any insights of the late nineteenth and early twentieth centuries are to be obtained, historians must

> deconstruct definitions of economic activity and family life that obscure the intimate relations between work, society, economics, and the family ... Until we redefine economics to include the variety of activities carried on to "make a living" within the household and redefine the family as an economic and political site – not simply an affective one – the culture and society of the nineteenth century rural majority will remain obscure, marginal to the "real" political and economic concerns of historians.[26]

There is much merit in this argument. Although I address some of Sandwell's concerns, I do not agree with her on all counts. For example, rather than examine politics as it affected family units (important as it is), I argue that one must first arrive at a better understanding of how such politics were conceptualized. Only then can one make assessments of how family dynamics influenced this construct.[27]

As Sandwell notes, the breadth of rural history, with its various subjects and diverse interpretative frameworks, means that no "school" of Canadian rural or agricultural history has emerged. John Herd Thompson, in writing on the state of rural/agricultural history, celebrates this diversity because it allows historians to explore numerous subfields that have yet to be adequately addressed, and because it prevents the onset of rigid orthodoxy. He argues, though, that one key element in rural life, "traditional" politics, has been all but ignored in recent work on Canada's rural past. In the main, historians have abdicated responsibility for the area "to political scientists and sociologists, who repeat the same questions about the CCF in Saskatchewan and Social Credit in Alberta before reaching the unhelpful answer that farmers behaved as they did because they were a homogeneous class of 'petit bourgeois independent commodity producers.'"[28]

Another problem for students of rural history in Ontario is that most research in the field is centred on Prairie farms and farmers or on rural life in Atlantic Canada.[29] Such work, while valuable, especially in helping historians come to terms with rural inequality and hierarchical social structures, also reinforces the notion that rural history is not an important field within Ontario historiography.

Most of the work undertaken in the rural history of Ontario consists of studies that either address specific communities of interest,[30] deal with specific localities,[31] or reassess the role agriculture has played in the provincial economy.[32] Rural women have received some attention of late,[33] and a few anecdotal accounts of rural history have been written in recent years.[34]

THE LIMITS OF CONVENTIONAL THEORY

The study of agrarian populist movements is plagued by the lack of theoretical perspectives in which to place one's analysis. Mass democratic movements are often referred to as "populist," but populism does not easily lend itself to definition, particularly since political activists on both the left and right have claimed ownership of the word. As John Richards notes, however, populism does have a meaning in everyday language: movements are populist "to the extent they display a strong faith in the 'common man's' virtues, in the ability of ordinary people to act together politically despite potentially serious class, racial, regional, or religious cleavages among them. Populist movements argue that concentrated political and economic institutions wield unwarranted power and, as a corollary, demand decentralization of economic and political power to the 'people.'"[35] Even so, populism resists any simple theoretical categorization.

Lawrence Goodwyn was confronted with this problem in his studies of the populist movement in the United States. Faced with working within one or the other of the two most prevalent historical paradigms in the 1970s (liberal capitalist and Marxist), he concluded that neither allowed for a satisfactory explanation of the movement. Consequently, he developed his own framework, which centred around a "movement culture" that placed the highest priority on people playing a meaningful role in how their lives were governed. His influence on this work is discussed below.

It is difficult to find examples of Marxist interpretations of farmers' movements in the Canadian context, since few Marxists consider farmers significant enough to merit serious examination.[36] Farmers, however, formed too great a percentage of the population to be completely ignored.[37]

There are two notable Marxist studies that have some bearing on the subject of this book. Russell Hann's work on the efforts of farmers to cope with an emerging industrial society in late-nineteenth-century Ontario is insightful, if for no other reason than that he takes agrarians on their own terms. He argues that, as "losers" in an increasingly urbanized society, farmers have been largely ignored. When one enters into their world, though, one can see "that there was widespread popular opposition to the creation of the Canadian commercial framework," an opposition that found expression in agrarian organizations. Equally important is Hann's belief that farmers "viewed the industrialization and urbanization of Canadian society from a perspective denied to those caught up in the logic of the process." They were thus able to develop a probing critique of what they saw[38] and put forward a vision, based on very different values, of how society should function. One may quarrel with Hann's preoccupation with industrialization (to the exclusion of much else that may have contributed to the farmers' militancy), but a debt is owed him for rescuing Ontario farmers from the indifference or condescension that is typical of most contemporary works. Because of its brevity, however, Hann's study only scratches the surface.

More recently, T. Robin Wylie evaluated the accomplishments of Ontario farm leader W.C. Good. Wylie's interpretation of the farmers' movement provides a good example of some of the debatable conclusions reached by those who approach agrarian political movements from a Marxist perspective. He begins by asserting that, historically, there have been two approaches within modern capitalism in search of direct democracy: populism and revolutionary socialism. He allows that both traditions recognize the need for change from below, and that both have been challenged by politics from above (that is, a reformist rather than radical leadership). He then notes that the populist approach (as tried in the UFO/Progressive movement) failed because the farm community was "unable to rise above its sectional interests," since ultimately farmers were "crypto-Liberals who wanted either fusion with the Liberals or a broad People's Party, or reformers from below who advocated a positive experiment in group government."[39]

The problem, argues Wylie, is that in the late nineteenth and early twentieth centuries, Canadian society was characterized "by a massive rural middle class, a powerful urban bourgeois, and a small and inexperienced working class." From their "intermediate class position," farmers could "look forward politically yet be economically reactionary." In short, in Wylie's view, an agrarian populist movement is democratic only in so far as democracy furthers the class ends of its members. Regarding economic issues, it is inherently undemocratic since it inevitably fights to preserve class position. Farmers who became adherents of agrarian radicalism were thus expressing "a sense of identity and loyalty to the dominant social order. But as threatened small property owners, they expressed this identification in a critical manner, by trying to reshape religion, patriotism, economic policies, and government reforms to suit their class needs." Wylie asserts that Canadian populists "were a notoriously self-centred group. They refused to look beyond agricultural issues … refused to cooperate with labour, and refused to take any positive direction from their leaders on whether to be organized as a party or in a positive group government fashion."[40]

These assertions are addressed in subsequent chapters, but a few words are in order here. First, Wylie's claim that there was an undifferentiated rural middle class borders on the nonsensical. As will be seen, early-twentieth-century rural Ontario was anything but homogeneous. Granted, some farmers were prosperous, but there were also many for whom each year meant a struggle merely to survive. As to farmers' self-centredness, it will be demonstrated that farmers often went far beyond agricultural issues and their immediate class concerns in their social criticism. Moreover, it will be shown that impressive efforts were made at the local level to secure unity between farmers and urban workers. Finally, it will be argued that it was the very act of farmers' listening to the movement's leaders that helped forestall their development of a more probing critique of the society in which they lived.[41]

Instead of relying on the interpretative frameworks described above, this work is informed by aspects of anarchist theory. Although numerous styles of thought and action have been referred to as anarchist,[42] for my purposes, anarchism will be defined as a "definite trend in the historical development, which ... strives for the free unhindered unfolding of all the individual and social forces in life ... For the anarchist, freedom is not an abstract philosophical concept, but the vital concrete possibility for every human being to bring to full development all the powers, capacities, and talent with which nature has endowed him, and turn them into social account."[43]

An anarchist approach provides a wider scope of analysis than a Marxist one. It allows, for example, for an exploration of the innate creative and humane traits in people; it helps make sense of their concerns for decentralization and local autonomy; it allows for a greater understanding of the cooperative impulse in humans;[44] it avoids grand theories; and it also holds that people have an instinctive aversion to power and authority. Marxism, with its concentration on class position and its belief in the impossibility that those with even a modicum of power *vis-à-vis* their control of the means of production would ever advance a radical thought, is obviously far more restrictive in this regard. More importantly, anarchism allows for a reconciliation between "individualism" and "collectivism," two seemingly oppositional theoretical perspectives, according to much of what has been written lately.[45]

None of this is to suggest that UFO members were proto-anarchists bent on overthrowing the state and all agencies of oppression or all manifestations of power. Yet if one accepts the main thrust of anarchist theory, which posits that there is a historical tendency (that finds more intense expression at certain times than at others) for people to question society's "truths" and to try to play a meaningful role in how their lives unfold, then one may approach the UFO with something other than condescension. In short, when viewed through the lens of anarchist theory, the thoughts and actions of UFO members make sense.[46]

OTHER INFLUENCES

In addition to anarchist theory, this work is informed by studies undertaken by Lawrence Goodwyn, Gregory Kealey, Bryan Palmer, James Naylor, and David Laycock.

Lawrence Goodwyn's ground-breaking work on late-nineteenth-century American populism did much to reconceptualize populist movements.[47] Refusing to accept the liberal capitalist paradigm of farmers' being history's losers and therefore unworthy of study, and the Marxist contention that farmers were incapable of radical thought given their class position, Goodwyn saw the populist movement as a group of people struggling to inject democracy into a polity that they perceived to be undemocratic.

Central to Goodwyn's analysis is the notion that populist movements represent the attainment of "individual self-respect" and "collective self-confidence," an attainment that permits people to conceive of the idea of acting in self-generated democratic ways – as distinct from participating passively in various hierarchical modes bequeathed by the received culture.[48] Goodwyn presents a sequential process by which a mass of people become a "movement culture" of opposition. Equally important, if not more so, he points out that "mass democratic movements are overarchingly difficult for human beings to generate." So much can go awry, and a "failure at any stage of the sequential process aborts or at the very least severely limits the growth of the popular movement."[49]

Related to this, Goodwyn's studies also touch upon the cultural "creation of mass modes of thought that literally make the need for major additional social changes difficult for the mass of the population to imagine."[50] This idea has importance here because it is argued that, despite great efforts to break free from prevailing attitudes, UFO members were never completely divorced from some of the fundamental assumptions that served as the underpinning for the official or hegemonic culture.

Kealey and Palmer, who studied workers in Toronto and Hamilton and collaborated on a history of the Knights of Labor, have done much to advance the fields of labour and social history. They have done so by looking beyond the institutional bodies that claimed to represent the interests of labour and by choosing instead to study workers' associations and the recreational life of workers, and to assess how this life manifested itself politically. Taking this approach, they have been able to conclude that there was something that could be termed a working-class culture and that this culture enabled workers to posit an alternative vision.[51] Central to their work is the idea that the lives of common people do matter and that, despite the onslaught of received culture, people are still able to mount challenges to the established order.

Another influence is James Naylor and his study of labour's challenge to the Ontario establishment during the period after the First World War. Beginning with the assertion that the class battles of that period were not fought exclusively in western Canada, Naylor argues that workers in Ontario were faced with the same challenges that workers in the West encountered, and that their response was equally important. No less significant is Naylor's assertion that labour's challenge during that period provides "an example of the creative potential of a workers' movement freed from the constraints of 'normalcy,' and free to dream of a new, more equitable, social order. Central to this notion of 'reconstruction' was a conception ... that the international struggle for democracy must be brought home."[52] Naylor's account is particularly sensitive to the remedies proposed by workers; he examines them within the context in which they emerged,[53] and he addresses some of the less-generous ideas that were developed.[54]

Naylor also explores the relations between organized labour and the UFO, and he shows that, in many respects, farmers and labourers had similar reasons for being disenchanted with the existing political parties and politics.[55] He demonstrates where farmers and workers differed on such matters as the eight-hour day but at the same time notes that in many ridings the UFO and the ILP worked in concert. In fact, in only two ridings in the 1919 provincial election did the UFO and ILP oppose one another.

Naylor concludes that if democracy is partly defined as "mass participation in deciding social priorities and direction," then the uprising in postwar Ontario "qualifies as a unique democratic moment in our history. It was an imperfect moment, however, as labourism had no ready solutions to workers' lack of power, and the undercurrent of nativism survived to return with a vengeance in the 1920s. Nevertheless, a massive, popular, debate on the nature and direction of Canadian society raised a hope of possible futures."[56] That the immediate postwar era represented a unique democratic moment in Ontario is supported here; that the motive power for this moment came primarily from the workers is not. In addition, although Naylor's point regarding the movement's lack of direction has merit, he does not factor in to any significant degree the role of hegemony in bringing workers and farmers back into the fold of "acceptable" political behaviour. In this book, an attempt has been made to do so.

Finally, the work of political scientist David Laycock has relevance to this study.[57] His examination of populism on the Prairies ranks among one of the more important works on mass democratic movements in Canada. Beginning with his observation that "Canadians have not always accepted determination of the legitimate scope and character of democratic politics by governments and élites," Laycock proceeds not only to show the richness and creativity of Prairie populist thought but also to take issue with the view that the class position of populists stalled any efforts at real change: "Being independent commodity producers did not circumscribe their political and social visions to the extent that they could entertain only paradigmatic, *petit bourgeois* notions and perspectives. In political life democratic thought is not parcelled up with the assignments of class copyrights to all of the elements that have constituted it over the years."[58] Laycock points out that there were distinct streams of populist thought, ranging in intensity from "crypto-Liberal" to "radical democratic." These categories are important if for no other reason than that they represent an attempt to address the numerous and diverse ideas from which individuals could choose.

Laycock also stresses the role of cooperation in pushing a populist movement towards greater democratic forms: "Participation in cooperative enterprises gave hard-pressed farmers self-respect and a sense of the democratic possibilities" that could sustain a movement.[59] He points out that the parallel rise of technocrats (accompanied by the growing belief amongst members of the public that experts could solve social, political and economic ills) created profound problems for the movement. What he does not do, however, is merge these themes so as to explain why the great cooperative ventures did not produce a sustained mass democratic movement. Such an effort is made here, because the demise of certain aspects of the cooperative movement in Ontario created problems for the UFO that, in the end, turned out to be insurmountable.

Equally significant for this book is Laycock's examination of the role of the state in populist conceptualizations of the "good society." In most cases populists believed that the state must be involved in eliminating the power of those who abuse it. More importantly, many populists recognized the positive uses of state power, and their recognition marks the moment, according to Laycock, "when the abstract objection to the state as a coercive force becomes vestigial in populist discourse."[60] This was certainly the case for the UFO, which, despite itself, attained political power in the province in 1919; before its 1919 victory the UFO had been a persistent critic of the role of the state in reducing farmers' democratic possibilities.[61]

Certainly, Laycock's work is not without problems. He tends, for example, to be rigid in his categories. Some people were crypto-Liberals, others agrarian radicals, and still others either social democrats or social credit adherents. This propensity for categorization is hard to avoid when trying to identify streams of populist thought. Such categories lose their meaning, however, when local farmers are examined. At times radical democratic impulses can be discerned, at other times there is clear evidence of crypto-Liberal thought. At all times, though, one sees people trying to understand a world that does not make sense to them.[62] Another troubling aspect of Laycock's work is that, as with virtually every other account of agrarian radicalism, he concentrates on the movement's leadership. Rank-and-file members, while often mentioned in general terms, are only called to the forefront to help underscore points already raised.

A further problematic aspect to Laycock's work is his position that Prairie populists were inherently more radical (or at least potentially more radical) than their counterparts in Ontario. On page after page the argument is made that it was on the Prairies that one witnessed "concerted and diverse attempts to reconstitute the democratic experience within the Canadian polity."[63] As for Ontario, the following passage sums up Laycock's perceptions: "Most UFO voters saw it as a moderate reform party for rural Ontarians and others interested in 'honest' government. Like ... the Patrons of Industry, the UFO was not seen by most as a fundamental alternative to either the established party system or the prevailing economic system."[64] Given this view, Laycock spends little time discussing populism in Ontario. Instead, he focuses on the Prairie brand of mass democratic action. As I shall argue, many of the traits that Laycock ascribes to the movement on the Prairies are also applicable in the context of Ontario. If it is the case that the attempts of Prairie populists to reform their society have been trivialized, then the efforts of Ontario populists have been marginalized to an even greater extent.

FINAL COMMENTS

This study examines the concerns of local UFO members, and the constant inundation of their vision of society as it should be by the world as it was. Key to understanding this inundation is the "propaganda system" of the official culture and its impact upon individuals and small groups. As mentioned, I examine that aspect of human nature that allows individuals to break free from orthodoxy and to conceptualize alternatives. At the same time, I shall describe the process by which hegemonic forces act to stifle this capacity. Throughout the book frequent reference will be made to the press, particularly rural newspapers. Such references are included to show how ideas articulated by the press shifted the terms of the debate significantly but so subtly that the shift was hardly noticed, much less commented on.

Much of this study is centred on elections, for two main reasons. First, naturally enough, elections were seen as important events. At the theoretical level they provided a rare occasion for citizens to take a meaningful hand in shaping their destinies. Hence they served as opportunities for people to express what mattered to them. Second, and more pragmatically, local newspapers devoted more attention to the UFO during elections than at other times. As a result, one gets the best impressions of the state of the movement (in terms of membership and in terms of what local clubs were up to) during elections.

Of course, one must never overlook the fact that the UFO meant different things to different people, and that the meaning consciously ascribed to it by members may differ from scholarly interpretations of that meaning. Monty Leigh, a Simcoe County farmer and rank-and-file member of the UFO, believed that the organization was formed because "farmers thought we should have more say in government."[65] For Leigh, the movement did not represent a break from orthodoxy, nor was it a manifestation of revolutionary potential: the UFO simply provided a mechanism through which farmers, who wanted to have more say in how their lives were governed, could attempt to effect change. This differs from the conventional assertion that the UFO was little more than a collection of cranky farmers showing their displeasure at the revocation of conscription exemptions, their concern over rural depopulation, their desire to make more money, or their fear of bigness and change. Of course, not all UFO members had the same goals and aspirations; some even used the UFO as a vehicle to maximize profits or to further other objectives.[66] That said, if one can speak of general trends within the movement, local UFO clubs exemplified people's desire to play a more substantial role in the unfolding of their lives and their relations with other individuals and groups.

Ultimately, the purpose of this book is to put names to a few of the thousands of Ontarians who saw things differently and who struggled to find meaning in a world that no longer made sense to them. They had witnessed a war that resulted in the slaughter of millions; they had seen first hand that the capitalists' promise of prosperity was largely empty (except for those few who had made the promises); and they had come to realize that the democracy for which so many had made sacrifices had not materialized. Much can be learned from their experience that has more than passing relevance today. Equally important, if not more so, this book is an attempt to restore to these people the dignity that they have long deserved but rarely been accorded. UFO members did not retreat into a shell; they did not respond in a scientifically predictable manner based on their class position; and they did not passively accept the hand that they had been dealt. Instead, they formulated a vision of society that challenged many fundamental assumptions. That they were able to create such a vision points to what can be accomplished when average citizens begin to conceptualize alternatives.

2 Historical and Statistical Profiles of Lambton, Simcoe, and Lanark Counties

The purpose of this chapter is to describe briefly (emphasizing physical and demographic characteristics) Lambton, Simcoe, and Lanark counties, to provide a preliminary sketch of who comprised the United Farmers of Ontario in these counties, and to offer some explanations for the UFO's composition.

As noted in chapter 1, most histories of the UFO either ignore the movement's composition – apart from the obvious observation that it was agrarian in nature – or, by focusing on its leadership, imply (without providing much evidence) that the UFO members represented a rural elite.[1] Since little effort has been made either to challenge or to substantiate these claims, determining the composition of the UFO is important. If it were shown, for instance, that the membership was drawn mainly from the landed gentry in the counties under consideration, then it could be argued that the movement amounted to little more than an assemblage of self-interested people who were concerned about the erosion of their class position. On the other hand, if members were drawn from the poorest of farmers in these counties, then it is possible to claim that the movement had the hallmarks of a radical insurgency of Ontario's underclass. If the movement consisted of farmers from all walks of life, various other possibilities present themselves. It might be argued, for instance, that the UFO was an inclusive organization that served to break down socioeconomic barriers and to enlist farmers from all sectors in a common cause. At the same time, however, the presence of certain prosperous and well-respected farmers in the membership may have forestalled any undertaking of a sustained critique of society.

The analysis that follows suggests that the third of these hypotheses gives the most accurate picture of UFO composition and character. The movement's membership in Lambton, Simcoe, and Lanark counties was, in the main, a mirror of the overall rural population in these localities. There were well-to-do farmers in the movement, and there were members who fought poverty at every turn. In most cases, however, the farmers who joined the UFO appear to have been the "average folk" of their communities.

Before describing and analysing the UFO membership, some prefatory remarks are in order. First, documentation regarding local UFO clubs is fragmentary; in particular, there is a dearth of local minute books and membership lists. In fact, for the three counties under consideration, I found only three UFO minute books – all generated by clubs in Simcoe County.[2] Hence it was necessary to find other sources to determine who the members of the movement were.

Newspapers are a useful source of information regarding club membership because UFO meetings were often covered by the local press. Usually only members of the club's executive were mentioned, however, a problem that is addressed below. Even so, there are some accounts in which rank-and-file members were named, and it has been possible to reconstruct at least partial membership lists for some clubs.

The general data presented here have been calculated from printed decennial Census of Canada reports. Some data from the annual reports of the Statistics Branch of the Ontario Department of Agriculture have been used as well, though they are less detailed than the federally generated ones. In this chapter, data from the 1921 census and, wherever possible, assessment records from 1920–22 have been privileged since they provide concrete evidence regarding the conditions in which farmers lived at the peak of the movement.

It should be stressed that the following is not a comprehensive demographic study of UFO membership in each of the three counties. I have made no substantial attempt, for instance, to trace the counties over time, nor have I searched extensively through property records, probated wills, and so on: such work would have entailed a separate research project and will have to await future studies. My purpose in this chapter is to delve as deeply as possible into a particular assortment of records to determine who comprised the UFO movement, and to situate the members in relation to non-member farmers and to society in general.

LAMBTON, SIMCOE, AND LANARK – A THUMBNAIL SKETCH

Historical Background

Owing to geographical and human influences, Lambton, Simcoe and Lanark counties developed in different ways. This section briefly touches upon their origins and growth as European settlements up to the early twentieth century.

European settlement in Lambton County did not begin in earnest until the early 1830s. Located in southwestern Ontario, Lambton had a population of 1,728 in 1834. By 1851, however, aided by immigration and some internal migration, it had attained a population of over ten thousand.[3] Later in the 1850s, oil was discovered in Lambton. The first wells were drilled in the early 1860s, oil refineries soon followed, and the early twentieth century witnessed the emergence of petrochemical industries and other large-scale commercial enterprises, including milling, salt, navigation, and construction companies. By 1891 the county's population had risen to 58,810; however, the opening of the West and a movement to industrial centres in the United States led to a decline and then gradual levelling off in Lambton's growth. It was not until after the Second World War that the population again reached 1891 figures. Nonetheless, thanks to its rich agricultural land and valuable oil fields, Lambton enjoyed a healthy agricultural/industrial base during the early twentieth century. Its main urban centre was (and is) Sarnia, which by 1914 had a population of roughly fourteen thousand. Several smaller communities also thrived – the towns of Petrolia, which had attained a population of 4,357 by 1891, and Forest, which by the early 1900s had a population of approximately two thousand.[4]

Although the site of French missionary efforts as early as the 1600s, Simcoe County (located in central Ontario) was not settled permanently by Europeans until the early 1800s. While relatively rich soils – particularly those found near the Holland Marsh in the southern portion of the county – facilitated substantial agricultural production, Simcoe also benefited from its position on the Great Lakes waterway, and the shipyards at Penetanguishene and Collingwood attest to the importance of Great Lakes commerce to Simcoe's economy. By the end of the nineteenth century there were several small villages in the county in addition to towns such as Barrie (with a population of some 6,000), Collingwood (5,500), Orillia (5,000), Midland (2,000), and Penetanguishene (2,000). In addition to small and, in some cases, large industries in these towns, there were a number of large-scale lumber operations in the county,[5] and the end of the century witnessed a rise in the number of tourists as well as cottagers from urban centres such as Toronto, who were drawn to the Georgian Bay and Lake Simcoe areas. Agriculture, however, continued to play an important role in Simcoe's economy.[6]

Lanark County, located in eastern Ontario, was initially settled by Europeans in the 1810s. By 1825 nearly eight thousand people resided there, and by mid-century the population had reached 27,317. By that time – thanks largely to the excellent water-power capabilities that the county offered industrialists – there had emerged the beginnings of what were to be substantial woollen industries.[7] In addition, lumbering played an important role in the county's development, as did the manufacture of iron products by such companies as the Findlay stove factory in Carleton Place and Frost and Woods's agricultural implements factory in Smiths Falls. Several other important industrial concerns, such as Walmpole Pharmaceuticals and Jergens, were established in Lanark County in the late nineteenth and early twentieth centuries.[8]

The Physical Setting

The three counties included the best and the worst of physical environments in "Old Ontario." Lambton is situated in a region containing mainly grey-brown Podzolic soil, some of the highest-quality agricultural land in the province. Simcoe County straddles the line between this grey-brown Podzolic soil and the less-arable and less-fertile brown Podzolic soil. In their studies of the physical characteristics of the province, Miller and Hoffman developed a soil-classification system that rated soil quality on a scale of one to seven (one being the highest- and seven the lowest-quality soil). According to their model, Lambton is the most fertile of the three counties under consideration, with some 85.7 percent of its soil belonging to the high-quality classes. Simcoe contains 54.6 percent of such soil while Lanark, clearly the worst of the three, contains only 14.5 percent.[9] Other features, such as average daily temperature, annual precipitation, growing degree days, start of the growing season, date of first frost, and so on, also worked in favour of Lambton and Simcoe inhabitants and against anyone trying to farm in Lanark.[10] The climatic and geological conditions limited the types of crops that could be grown in Lanark and, equally important, the amount and quality of crops that could be produced.

Table 1
Rural Land Cleared in Lambton, Simcoe, and Lanark Counties, 1920

County	*Acres Assessed*	*Acres Cleared*	*%Cleared*
Lambton	658,445	525,598	79.82
Simcoe	967,478	687,471	71.06
Lanark	670,969	336,764	50.19

Source: Ontario, Department of Agriculture, *Annual Report of the Statistics Branch 1920*. Calculations by author.
Note: Figures for unorganized townships are not included.

Statistics generated by the Ontario government on land usage bear out these assertions. As shown in Table 1, the percentage of rural land cleared in Lanark was noticeably lower than that in the other two counties. Although large percentages of land had been cleared in Lambton and Simcoe by 1920, barely half of Lanark's acreage had been improved. Certainly, the poor quality of the soil and its limited agricultural potential discouraged clearing on the scale witnessed in the other two counties.

Demographic Features

In 1921 Lambton's total population was 52,102, of which 27,283 (fifty-two percent) were classified rural and 24,819 (forty-eight percent) urban. In 1921 Simcoe had a population of 84,032, of which 43,154 (fifty-one percent) were rural and 40,878 (forty-nine percent) urban. Lanark counted 32,993 residents, 15,549 (forty-seven percent) of whom were considered rural and 17,444 (fifty-three percent) urban.[11] These totals remained fairly stable during the early twentieth century, although Lambton and Lanark had experienced a slight decrease in population from the late nineteenth century. Only Simcoe recorded a population increase over the 1911–21 period.

When urban and rural populations in the three counties are separated, a different picture emerges with respect to population change. As seen in Table 2, virtually every township (i.e., rural area) in Lambton, Simcoe, and Lanark counties experienced a real decline in population between 1911 and 1931. In some townships, such as Enniskillen in Lambton, Tay in Simcoe, and Bathurst in Lanark, the decline was substantial. It will be seen that the UFO harboured negative feelings towards urban areas and urban society in general. It is reasonable to assume that rural depopulation in the counties fed into that animosity.

Table 2
Population Change in Lambton, Simcoe, and Lanark County Townships, 1911–31

Township	*1911*	*1921*	*1931*	+/–
LAMBTON				
Dawn	2,930	2,470	2,294	–636
Moore	3,771	3,611	3,381	–390
Plympton	3,206	2,829	2,695	–511
Sarnia	2,663	2,538	3,441	+778
Sombra		3,274	2,971	–303
Bosanquet	2,491	2,332	2,200	–291
Brooke	2,992	2,703	2,373	–619
Enniskillen	3,632	3,063	2,692	–940
Euphemia	1,846	1,543	1,455	–391
Warwick	2,772	2,376	2,234	–538
SIMCOE				
Matchedash	459	507	480	+21
Medonte	3,361	2,723	2,533	–828
Orillia	4,508	4,561	4,979	+471
Tay	5,245	3,159	2,770	–2,475
Tiny	4,121	4,026	3,693	–428
Flos	3,239	3,034	2,929	–310
Nottawasaga	4,432	4,110	3,759	–673
Oro	3,485	3,098	2,842	–643
Sunnidale	2,175	2,070	2,013	–162
Vespra	2,596	2,281	2,486	–110
LANARK				
Dalhousie	1,225	1,047	930	–295
Darling	623	441	424	–199
Lanark	1,595	1,289	1,247	–348
Lavant	547	429	407	–140
Pakenham	1,605	1,518	1,384	–221
Ramsay	1,942	1,713	1,679	–263
Sherbrooke N.	277	283	281	+4
Bathurst	2,228	2,023	1,789	–439
Beckwith	1,402	1,221	1,221	–181
Burgess N.	752	660	592	–160
Drummond	1,764	1,584	1,426	–338
Elmsley N.	1,066	779	788	–278
Montague	1,839	1,792	1,642	–197
Sherbrooke S.	742	770	718	–24

Source: Census of Canada, 1911–1931. Calculations by author.

Table 3
Ethnicity in Lambton, Simcoe, and Lanark, 1921

Ethnicity	*Lambton*	*Simcoe*	*Lanark*
English	23,154	33,564	7,218
Irish	12,500	23,468	13,575
Scottish	11,683	14,148	10,047
French	1,367	7,992	1,140
Other European	1,805	3,643	775
Asiatic	102	121	90
Aboriginal	541	320	18

Source: Census of Canada, 1921
Note: The numbers in the table do not add up to the total county populations because "not specified" are not included.

In all three counties, the population was overwhelmingly Canadian-born. Of the 52,102 people in Lambton, 44,745 (eighty-six percent) were born in Canada, 4,941 (nine percent) were British-born, and 2,416 (five percent) were listed as "foreign-born." In Simcoe, 74,457 (eighty-nine percent) of the total population of 84,032 were born in Canada, 8,386 (nine percent) were British-born, and 1,189 (two percent) were born elsewhere. In Lanark, of its population of 32,993, 30,211 (ninety-two percent) were Canadian-born, 2,145 (six percent) were British-born, and 637 (two percent) were foreign-born. Similarly consistent patterns emerge with regard to ethnicity. As Table 3 indicates, the overwhelming majority of people listed either England, Scotland, or Ireland as the country of their families' origin.

Considerable homogeneity is found in terms of religious affiliation as well. As seen in Tables 4.1 to 4.3, there were few denominations with more than two hundred members in each county. In Lambton, the largest denominations were Methodist, Presbyterian, and Anglican, along with sizable congregations of Roman Catholics and Baptists. In Simcoe, Methodists and Presbyterians comprised 27 percent and 26 percent of the total population respectively, followed by Anglicans (20 percent) and Roman Catholics (17 percent). In Lanark, Presbyterians formed the largest denomination, followed in turn by Anglicans, Methodists, and Roman Catholics.

There were 6,775 occupied farms in Lambton in 1921, as compared to 7,914 in Simcoe and 2,896 in Lanark. What is of more interest, however, is the size of the farms. As Table 5 indicates, most farms in each of the three counties were between fifty-one and two hundred acres in size. Only in Lanark was there a significant percentage of

Table 4.1
Religious Denominations with 200 or More Adherents, Lambton County, 1921

Religion	*Adherents*	*% of Population*
Methodist	20,103	34
Presbyterian	15,714	28
Anglican	10,389	20
Roman Catholic	4,791	9
Baptist	4,611	8
Congregationalist	727	1
Mormon	532	1
Salvation Army	421	1
Plymouth Brethren	200	.3

Table 4.2
Religious Denominations with 200 or More Adherents, Simcoe County, 1921

Religion	*Adherents*	*% of Population*
Methodist	22,719	27
Presbyterian	22,073	26
Anglican	17,341	20
Roman Catholic	14,797	17
Baptist	2,957	3
Brethren	596	.5
Salvation Army	562	.5
Mennonite	417	.5
Congregationalist	382	.5
Jewish	306	.5

Table 4.3
Religious Denominations with 200 or More Adherents, Lanark County, 1921

Religion	*Adherents*	*% of Population*
Presbyterian	11,191	34
Anglican	7,825	23
Methodist	5,809	17
Roman Catholic	5,665	17
Baptist	1,370	4
Congregationalist	395	1

Source: Census of Canada, 1921. Calculations by author.

Table 5
Farm Size by County, 1921

Acreage	*Lambton*	*Simcoe*	*Lanark*
1–50	2,079 (31%)	1,948 (25%)	248 (9%)
51–100	2,989 (44%)	3,477 (44%)	681 (24%)
101–200	1,443 (21%)	2,037 (25%)	1,066 (36%)
200+	264 (4%)	452 (6%)	901 (31%)
Total	6,775 (100%)	7,914 (100%)	2,896 (100%)

Source: Census of Canada, 1921. Calculations by author.

Table 6
Farm Size by County, 1931

Acreage	*Lamton*	*Simcoe*	*Lanark*
1–50	1,730 (27%)	1,729 (23%)	203 (7%)
51–100	2,721 (43%)	3,156 (42%)	622 (23%)
101–200	1,580 (25%)	2,160 (28%)	1,032 (38%)
200+	320 (5%)	546 (7%)	872 (32%)
Total	6,351 (100%)	2,729 (100%)	7,591 (100%)

Source: Census of Canada, 1931. Calculations by author.

farms of over two hundred acres. Although by 1931 there was a slight decline in the number of small farms in Lambton, Simcoe, and Lanark (as well as a slight decline in the number of farms in these counties), the figures remain fairly constant (Table 6).

A large farm does not, however, always ensure prosperity. Land must be improved if it is to be productive and profitable. In Lambton, of the 650,970 farm acres in 1921, 461,848 (seventy-one percent) were improved; in Simcoe in that year, of the 844,617 acres of farmland, 556,227 (sixty-six percent) were improved; in Lanark, of the 563,256 acres of farmland in 1921, only 195,785 (thirty-five percent) were improved. In Lanark, then, having a large farm might mean having large areas of woodland, rock, slash, or marsh.[12]

Farm tenancy, although certainly not unheard of, was fairly rare in all three counties during the period under consideration. Of the 6,775 occupied farms in Lambton, 5,850 (eighty-six percent) were owned by the occupants. Of the remaining farms, 478 (seven percent) were operated by tenant farmers, and 447 (seven percent) were occupied by part owners/part tenants. In Simcoe, of the 7,914 occupied farms, 6,683 (eighty-four percent) were operated by owners, 852 (eleven percent) were run by tenants, and 379 (five percent) were operated by part owners/part tenants. In Lanark, of the 2,896 occupied farms,

2,641 (ninety-one percent) were run by owners, 136 (five percent) by tenants, and 119 (four percent) by part owners/part tenants.

Owning a farm does not necessarily define an individual's financial position or, by extension, socio-economic status in the community. Indebtedness may provide a more accurate measure, but it is difficult to establish the amount of mortgage indebtedness for the three counties for the early 1920s, since it was not included in the 1921 census. Such figures do exist for 1931, however, and they indicate that a large percentage of farms carried mortgages. Of the 4,516 farms in Lambton County, 2,443 (fifty-four percent) indicated mortgage indebtedness. In Simcoe, of the 5,971 farms that provided information on the subject, 3,109 (fifty-two percent) reported such debts. In Lanark, of the 1,768 farms, 918 (fifty-two percent) carried mortgages. Thus, while it is true that many farms were of the "fully owned" status, it remains the case that even in these fairly long settled areas a considerable amount of mortgage indebtedness persisted – over fifty percent of the farms in each county carried mortgages.[13] It is reasonable to assume that similar patterns existed in the early 1920s.

Agricultural Production

Agricultural production is a telling story, not only in terms of crops grown in the three counties but also in terms of the money each of these crops brought farmers in those localities. In 1920, as Tables 7.1 to 7.3 indicate, the field crop that brought in the highest dollar amounts in Lambton and Simcoe was wheat, followed by oats in both counties. In Lanark, not known for its productive soil, the chief crop in terms of dollar amounts was hay, followed by oats. In all three counties, grain and forage crops made up a significant percentage of produce, although Simcoe farmers grew substantial quantities of root crops.

What is of even greater interest is land productivity as shown in the three tables. Lanark farmers tended to produce fewer crops per acre than their counterparts in the other counties. In the case of wheat, for example, Simcoe farmers produced an average of 22.4 bushels per acre, while Lambton farmers produced an average of 21.4 bushels per acre. In Lanark, however, the average return was 16.3 bushels per acre, markedly lower than the yield in the other two counties.

In 1920 local markets still reflected local conditions;[14] if a crop was not produced in abundance in a given area, prices tended to be higher than in areas in which the crop was produced in greater quantities. Thus, in some cases, Lanark farmers were able to realize

Table 7.1
Field Crops – Acreage, Yield, Value, and Return per Acre, Lambton County, 1920

Crop	*Acreage*	*Yield*	*$Value*	*Ret/Acre*	*$Ret/Acre*
Cultivated Hay	84,216	99,603t	1,497,903	1.2t	17.79
Oats	76,874	3,383,338b	1,627,558	44.0b	21.17
Wheat	58,007	1,241,200b	2,271,095	21.4b	39.15
Forage Corn	16,956	130,409t	564,972	7.7t	33.32
Husking Corn	13,561	764,632b	518,890	56.4b	38.26
Barley	9,188	301,307b	228,625	32.8b	24.88
Mixed Grains	6,905	290,992b	188,973	42.1b	27.37
Root Crops	6,060	63,911t	684,965	10.5t	113.03
Potatoes	2,148	256,445b	210,943	199.4b	98.20
Flax (Fibre)	885	193,840l	20,270	219.0l	22.90
Forage Crops	716	1,283t	11,742	1.8t	16.40

Table 7.2
Field Crops – Acreage, Yield, Value, and Return per Acre, Simcoe County, 1920

Crop	*Acreage*	*Yield*	*$Value*	*Ret/Acre*	*$Ret/Acre*
Oats	130,371	3,958,460b	2,031,157	30.4b	15.58
Cultivated Hay	95,298	107,528t	2,006,645	1.1t	21.05
Wheat	56,457	1,265,307b	2,370,340	22.4b	41.98
Barley	36,455	998,260b	821,276	27.4b	22.53
Mixed Grains	23,296	668,705b	483,519	28.7b	20.75
Rye	11,292	151,623b	200,922	13.4b	17.79
Potatoes	10,878	1,214,622b	686,607	111.6b	63.12
Forage Corn	10,378	82,071t	353,530	7.9t	34.06
Buckwheat	10,337	176,011b	171,886	17.0b	16.63
Peas	5,104	95,320b	153,780	18.7b	30.13
Turnips	4,014	1,634,332b	223,935	407.0b	55.79
Root Crops	1,736	17,060t	90,355	9.8t	52.04
Grains for Hay	1,180	1,438t	22,740	1.2t	19.27
Forage Crops	1,055	1,508t	18,905	1.4t	17.91

slightly higher dollar returns per unit produced (see Table 8). This was particularly true with respect to forage crops and some grains, notably barley and oats. In other cases, where the relatively small amounts of crop produced should have equated with higher dollar returns per unit produced, Lanark farmers received comparable returns for their produce relative to their counterparts in Lambton and Simcoe and, in some cases, even less. In these instances, one must assume that there was either little local demand for these crops or, more plausibly, that they were of an inferior grade.

Table 7.3
Field Crops – Acreage, Yield, Value, and Returns per Acre, Lanark County, 1920

Crop	*Acreage*	*Yield*	*$Value*	*Ret/Acre*	*$Ret/Acre*
Cultivated Hay	68,670	67,153t	1,368,296	.9t	19.92
Oats	41,849	1,215,813b	760,060	29.1b	18.16
Wheat	8,121	132,778b	239,522	16.3b	29.49
Forage Corn	7,671	64,385t	291,899	8.4t	38.05
Mixed Grains	5,891	173,675b	140,776	29.5b	23.89
Barley	4,084	108,821b	107,300	26.6b	26.27
Buckwheat	3,820	58,308b	61,945	15.3b	16.22
Potatoes	2,327	229,512b	153,636	98.6b	66.02
Grains for Hay	1,449	1,779t	26,558	1.2t	18.32
Forage Crops	916	1,072t	13,935	1.2t	15.21
Peas	725	11,284b	19,272	15.6b	26.58
Rye	599	8,458b	10,120	14.1b	16.89

Source: Census of Canada, 1921. Calculations by author.
t = tons b = bushels l = pounds

Table 8
Dollar Returns per Unit Produced – Selected Field Crops, Lambton, Lanark, and Simcoe Counties, 1920

Crop	*Lambton*	*Lanark*	*Simcoe*
Hay (cultivated) ($)	15.04/t	20.38/t	18.66/t
Oats ($)	0.48/b	0.62/b	0.51/b
Wheat ($)	1.83/b	1.80/b	1.87/b
Corn (forage) ($)	4.33/t	4.53/t	4.31/t
Barley ($)	0.75/b	0.98/b	0.82/b
Mixed Grains ($)	0.64/b	0.81/b	0.72/b
Potatoes ($)	0.82/b	0.67/b	0.57/b
Other Forage Crops ($)	9.15/t	13.00/t	12.54/t

Source: Census of Canada, 1921. Calculations by author.
t = tons b = bushels

Even when there were higher returns per unit produced in Lanark, any benefit was negated by the fact that Lanark farmers, on average, produced much less per farm than their counterparts in Lambton or Simcoe (see Tables 7.1 to 7.3). Clearly, then, Lanark lagged behind Lambton and Simcoe in terms of individual farm income, at least as far as field crops were concerned.

In terms of fruit production, one finds a fairly consistent return per unit produced (Table 9), albeit with minor variations resulting, presumably, from local market prices. There are differences, however, in the amount and type of fruit grown in each county. Lambton shows

Table 9
Fruit Production – Lambton, Simcoe, and Lanark Counties, 1920

Fruit	*Lambton*	*Simcoe*	*Lanark*
APPLES			
Produced	600,825b	266,185b	21,021b
Value ($)	407,574	100,927	21,738
Ret/Unit ($)	0.68	0.68	1.03
PEACHES			
Produced	21,059b	14	–
Value ($)	27,532	13	–
Ret/Unit ($)	1.31	0.93	
PEARS			
Produced	12,260b	1,452b	–
Value ($)	13,043	2,088	–
Ret/Unit ($)	1.06	1.44	
PLUMS/PRUNES			
Produced	9,942b	4,099b	252b
Value ($)	11,686	6,240	391
Ret/Unit ($)	1.18	1.52	1.55
CHERRIES			
Produced	3,216b	2,664b	19b
Value ($)	5,819	5,457	32
Ret/Unit ($)	1.81	2.05	1.68
GRAPES			
Produced	48,932l	2,228l	1,218l
Value ($)	3,425	156	85
Ret/Unit ($)	0.07	0.07	0.07
STRAWBERRIES			
Produced	121,940qu	236,615qu	12,806qu
Value ($)	23,563	42,993	2,981
Ret/Unit ($)	0.19	0.18	0.23
RASPBERRIES			
Produced	147,446qu	157,646qu	1,320qu
Value ($)	32,806	28,938	293
Ret/Unit ($)	0.22	0.18	0.22
CURRANTS			
Produced	30,114qu	16,157qu	1,156qu
Value ($)	5,950	3,055	216
Ret/Unit ($)	0.20	0.19	0.19
GOOSEBERRIES			
Produced	1,673qu	10,961qu	2,021qu
Value ($)	197	1,552	303
Ret/Unit ($)	0.12	0.14	0.15

b = bushel l = pound qu = quart

Table 10
Livestock – Lambton, Simcoe, and Lanark Counties, 1921

Stock	*Lambton*	*Simcoe*	*Lanark*
HORSES			
Number	21,973	29,409	10,337
Value ($)	2,604,937	3,773,977	1,236,102
CATTLE			
Number	90,855	104,891	58,907
Value ($)	4,414,559	3,833,567	1,952,031
SHEEP			
Number	24,060	52,949	34,881
Value ($)	170,279	346,283	224,123
SWINE			
Number	52,165	65,284	17,192
Value ($)	496,223	618,963	142,649
HENS AND CHICKENS			
Number	612,322	554,159	209,553
Value ($)	322,537	323,092	117,594
TURKEYS			
Number	29,766	11,938	6,062
Value ($)	29,303	15,512	5,716
DUCKS			
Number	19,380	10,842	1,366
Value ($)	10,515	8,430	956
GEESE			
Number	13,209	19,467	1,946
Value ($)	16,648	25,962	2,956

Source: Census of Canada, 1921.

strong production figures for virtually all fruits listed in the census. Simcoe County farmers, with cooler summers, a shorter growing season, and somewhat poorer soil than Lambton farmers enjoyed, produced relatively small amounts of some fruits. In Lanark, with even worse climatic and geological conditions, production of certain fruits was insignificant, where it occurred at all.

The one area in which Lanark farmers seem to have held their own was livestock. Despite its relatively small population, the county had strong figures in certain types of livestock, as Table 10 shows. In particular, cattle and sheep were prominent in many Lanark farmers' operations. These figures are not surprising: given the county's relatively poor soil quality and climatic conditions, it made sense for farmers to turn to livestock for milk, wool, and meat. As will be seen, the poultry, sheep, cattle, and swine holdings of farmers in all three counties – all of which were held in sufficient quantities to produce

Table 11
Maple Syrup and Sugar Production, Lambton, Simcoe, and Lanark Counties, 1920

County	*Syrup*	*Sugar*	*Total Value ($)*
Lambton	8,294gal	42lbs	24,545
Simcoe	12,701gal	675lbs	30,592
Lanark	48,243gal	5,706lbs	98,499

Source: Census of Canada, 1921.

for non-farm consumption – go some way towards explaining the growth in cooperative marketing activities (both the establishment of new associations and the augmentation of those already in existence) that accompanied the rise of the UFO.

There were, of course, other ways to augment farm income. Wood products (firewood, staves, shingles, fenceposts, rails, potash, pulpwood, etc.) continued to contribute significantly to farm income in some areas well into the twentieth century. In addition, farmers could obtain extra cash by providing labour and animal power for transport services and for construction.[15] Other money-making farm activities included the curing of meats and the spinning and weaving of wool, and, as Table 11 suggests, if a farmer was fortunate enough to possess sugar maples, then some cash could be obtained through the sale of maple syrup products. In Lanark in particular, such sales could add several dollars to a farm's total income.

To obtain an estimate of average annual farm income in each of the three counties, the income from all farm activities (as listed in the 1921 Census of Canada) has been totalled and this figure has then been divided by the number of occupied farms in each county (Table 12).

Based on these figures it is apparent that, on average, farms in each county produced crops and other products worth roughly the same value. The numbers change, however, when total farm value, as determined in the 1921 Census of Canada, is considered. The total farm value is determined by adding the values in Table 12, and by then adding to this figure the values of land, buildings, implements and machinery, and livestock. As Table 13 indicates, Lanark lagged behind Simcoe by roughly $1,000 on average, and behind Lambton by over $1,500. In many cases, farmers fell well short of these averages while other, more fortunate ones undoubtedly exceeded them.

Another way to obtain an idea of the relative prosperity of farmers in each county is to examine the value of farm property, implements, and livestock on hand – in other words, farm assets. When farm

Table 12
Value of Farm Products – Lambton, Lanark, and Simcoe Counties, 1920

	Lambton	*Simcoe*	*Lanark*
PRODUCT			
Field Crops ($) (a)	7,978,427	10,334,851	3,297,925
Fruits ($) (b)	557,228	303,085	123,507
Forest Products ($)	49,894	175,166	177,901
Stock Sold Alive ($)	3,240,133	3,569,297	1,271,655
Animal Products ($) (c)	2,752,906	3,278,287	1,761,677
Total ($)	14,578,588	17,660,686	6,632,665
AVERAGE VALUE OF PRODUCTS PER FARM			
No. Occupied Farms	6,775	7,914	2,896
Avg. per Farm ($)	2,151.82	2,231.58	2,290.28

Source: Census of Canada, 1921. Calculations by author.
(a) Includes produce consumed on farm as well as crops sold.
(b) Includes maple syrup.
(c) Includes dairy products, eggs, honey, wax, and wool.

Table 13
Total Farm Values, Lambton, Simcoe and Lanark Counties, 1920

	Lambton	*Simcoe*	*Lanark*
Total Value ($)	59,405,383	63,806,525	20,450,610
Avg. per Farm ($)	8,768.32	8,062.49	7,061.67

Source: Census of Canada, 1921. Calculations by author.

assets are added and an average is determined for each county (see Table 14), one sees again that, on average, Lanark farmers were less well off that their counterparts in Lambton and Simcoe; on average, their farm assets were worth at least five hundred dollars less than those of Simcoe farmers, and at least seven hundred dollars less than those of Lambton farmers.

Clearly, there were differences in farming in the three counties. This is particularly evident when the figures for field crops, fruit, and livestock are calculated per farm in each county. For instance, there were, on average, 13.4 head of cattle per farm in Lambton, compared to 13.2 in Simcoe and 20.3 in Lanark; Lambton farmers each produced, on average, 88.7 bushels of apples, compared to 33.6 in Simcoe and 7.3 in Lanark; and Lanark farmers planted on average twice as much hay as the other two counties. As the preceding tables show, Lambton farmers were heavily engaged in fruit farming and also specialized in fowl. Simcoe represented a typical mixed-farming

Table 14
Farm Property, Implements, and Livestock, Lambton, Simcoe, and Lakarn Counties, 1920

	Lambton	*Simcoe*	*Lanark*
Land ($)	32,485,525	33,705,588	11,628,748
Buildings ($)	15,648,642	20,714,518	5,982,410
Implements ($)	5,513,203	6,310,410	2,071,385
Livestock	11,889,968	14,092,887	6,226,890
Total	65;537,338	74,823,403	25,909,433

AVERAGE VALUE PER FARM:

County	*No. Farms*	*Avg. per Farm ($)*
Lambton	6,775	9,673.41
Simcoe	7,914	9,454.56
Lanark	2,896	8,946.63

Source: Census of Canada, 1921; Ontario, Department of Agriculture, *Annual Report of the Statistics Branch 1920.* Calculations by author.

area, with middling numbers in almost all categories. Lanark farmers, conversely, specialized in dairy cattle, sheep, and pasture.

These figures invite some speculation. Clearly, the UFO attracted several types of farmers – commercial agrarians as well as "average" and marginal farmers. Although this speaks to the inclusiveness of the movement, the differing motives of the members and the different needs of each type of farmer may also have helped pull the UFO apart as a political movement. As I argue in subsequent chapters, the slightly greater militancy in Lanark relative to Lambton or Simcoe may have been due, in part, to the county's less-diversified and less-prosperous rural economy. Discrepancies between the counties overall may also explain, at least partly, why Lambton, with its greater level of professionalism and affluence, adhered more strongly to the principle of group government (i.e., electing legislators based on occupational groups in proportion to their numbers in society) rather than to the notion of broadening out the movement to include other, non-farm groups.

COMPOSITION AND CHARACTER OF UFO MEMBERSHIP

Methodology

The foregoing shows that, despite certain physical, demographic, and productive similarities, each of the three counties was to some degree distinct. Were the differences among them reflected in each county's UFO membership? To answer this question, we turn to township assessment rolls and to assembled UFO membership lists.

In order to determine who joined the UFO in Lambton and Lanark, names were culled from local newspaper accounts of the UFO for the period 1916 to 1930. If it was not clear whether a person was a UFO member, their name was not used. The remaining names were then matched up with the pertinent entries in assessment records for selected townships in each county. In Lambton, all but Sombra, Dawn, and Sarnia townships were analyzed. In Lanark, the townships of Darling, Dalhousie, Lavant, South and North Sherbrooke, Pakenham, North Burgess, and Bathurst were not thoroughly analyzed because they contained too few UFO members.

Matching up UFO members against thousands of names on assessment rolls is a highly time consuming task. Thus, in the case of Simcoe County, where there are records of three local farmers' clubs, I decided to focus only on the townships where these clubs were located. In effect, the Oro Station and Rugby UFO clubs in Oro Township and the Edenvale UFO Club in Flos Township can be seen as a sort of "control group" *vis-à-vis* the other counties. What follows should by no means be considered an exhaustive listing of UFO members in these counties, but it should certainly be viewed as suggestive.

As noted earlier, using newspaper accounts to locate UFO members presents certain problems. One of the main concerns is that many of those referred to in newspaper stories served, in some capacity, on the executives of local clubs and might have enjoyed a higher socioeconomic status than rank-and-file members. Fortunately, the three extant minute books from Simcoe provide lists of members elected to executive positions for each year, and these lists can be used to determine whether local executive members were of a different socioeconomic class from average members.

Assessment records are a rich source of information about individuals; they often include the age of the property owner, number of acres owned, number of acres cleared, amount of non-productive land owned,[16] in some cases the religious denomination of the property owner, and whether the occupant owned or leased the land.[17] That said, a number of caveats should be mentioned. First, although the rolls show what the assessor considered to be the value of property, specifically buildings and land, there was often no consistency in the evaluations made from one county to another or, for that matter, from one township to another. This is not too great a problem for my purposes since I am interested not in the intrinsic value of property but only in the *relative* value of UFO members' property as compared to that of non-members in the same locality.

Second, assessors did not always record all of the information requested on the preprinted forms. Thus, in some townships the religious affiliation of property owners was dutifully identified, in others it was not. The same applies to the ages of owners and to other matters that are addressed below.

Assessment records present additional problems for those attempting to uncover data regarding the members of an organization. In each of the three counties, some UFO members could not be located, despite thorough and repeated efforts to do so. There may be several reasons for this. First, many member farmers may have found it more convenient to join clubs in a different county, especially if they lived close to county boundaries. This is certainly true in areas such as Smiths Falls, where, for example, the long-serving Smiths Falls UFO club president, John Willoughby, lived in neighbouring Leeds-Grenville. Time and other constraints did not allow for searches to be undertaken in the assessment rolls of other counties.

Many of the members may have been farmers' sons, whose names did not always appear in assessment records. There were also several instances in which farmers' daughters were named as members in newspaper accounts, but in such cases it was impossible to match the person to the parent who would have been listed in the assessment roll.

Third, owing to research constraints, I decided to use only one assessment roll per township. I tried to choose rolls dating from around 1920 because the information they contain corresponds most closely with the 1921 Census of Canada data. The problem is, however, that a UFO member who left the area in 1919, or a farmer who moved to the area in 1925, would not be listed in a 1920 assessment roll.

Fourth, in many newspaper accounts of UFO meetings, first initials were used instead of Christian names. Occasionally, moreover, a member was identified only as "Mr Jones." Again, it was not always possible to match these individuals, especially when the designation was "J. Wilson" or "D. Smith," to names on the assessment rolls. It seemed best, in such cases, to err on the side of caution and leave these names out, rather than select a corresponding name at random from the assessment roll merely to have it included in the tables.

Finally, anyone who uses assessment records for statistical purposes invariably encounters the problem of multiple ownership of property. In some instances, two or more individuals were listed as owning one parcel of land; in others, two or more family members were listed as owners. The latter case is easier to deal with: whenever a farmer and what is clearly his son are listed as joint owners, the entire value of the property has been included in the calculations. A determination was made for inclusion based on the assumption that the entire property would pass to that son on the death of the parent. In cases where two brothers co-owned a parcel of land, the value has been halved to reflect the wealth of each brother more accurately. In instances of co-ownership by what seem to be two unrelated people, the property value has been left as is, and a notation has been made.

Some further explanatory notes regarding methodology are in order. The method used for determining the socioeconomic status of UFO members is fairly straightforward. Known UFO members have been matched with the entries pertaining to them in township assessment rolls. The numerical values ascribed to these people have been totalled (when appropriate) and then averaged out.

The results of these calculations, while providing some insights regarding UFO membership, do not reveal where members stood relative to other farmers in the townships in which they lived. In order to compile a manageable control group of non-members, a random selection was assembled for each locality by taking the first three farmer property owners from each page of the township assessment rolls. The selection of three from each page was designed to provide a sampling of farmers from all areas of the township (known UFO members were themselves located in most parts of a given township) and to prevent unintentional focusing on particularly wealthy or poor areas. The numerical values associated with these farmers were totalled and then averaged out. In cases where farmers were listed as tenants, or where there was some doubt as to whether or not the person enumerated was a farmer, the individuals were ignored in the control group, and the next farmer was selected.[18]

Some of the people in the control group may also have been UFO members. But they were also residents of the township, and the purpose here is to provide a general local picture, not an in-depth study of the property-owning dynamics of each township.

Distinguishing Features of UFO Membership by County

Based on the data obtained through the survey of local assessment records, the overall county averages for Lambton and Lanark counties, and the averages for Oro and Flos townships, Simcoe County, are as follows (Table 15). Keeping in mind that these are aggregate averages, the results are nonetheless intriguing inasmuch as similar patterns are exhibited in each county.

Table 15
UFO and Non-UFO Membership – Lambton and Lanark Counties and Oro and Flos Townships, Simcoe County

	UFO	*Non-UFO*
LAMBTON COUNTY		
Number Identified	185	627
Avg. Age	44.74	47.94
Avg. Acres Owned	113.75	107.92
Avg. Acres Cleared	98.58	92.40
Avg. Acres Non-Prod.	15.17	15.52
Avg. Value Land ($)	3,551.35	3,488.97
Avg. Value Buildings ($)	692.45	687.09
Avg. Total Value ($)	4,243.80	4,176.06
LANARK COUNTY		
Number Identified	137	541
Avg. Age	46.06	48.63
Avg. Acres Owned	197.82	190.11
Avg. Acres Cleared	108.34	116.33
Avg. Acres Non-Prod.	89.48	73.78
Avg. Value Land ($)	2,260.60	2,382.62
Avg. Value Buildings ($)	740.00	659.96
Avg. Total Value ($)	3,000.60	3,042.58
ORO AND FLOS TOWNSHIPS – SIMCOE COUNTY		
Number Identified	80	216
Avg. Age	39.60	47.26
Avg. Acres Owned	142.51	135.30
Avg. Acres Cleared	102.27	93.88
Avg. Acres Non-Prod.	40.24	41.42
Avg. Value Land ($)	2,694.31	2,844.21
Avg. Value Buildings ($)	809.10	800.84
Avg. Total Value ($)	3,503.41	3,645.05

Source: AO, Assessment rolls; LCA, Simcoe County Archives, UFO minute books; various newspapers. Calculations by author.

The breakdown of known UFO members in Lambton by township is included in Appendices A to G. In townships with fifteen or more UFO members, the members seem to have been fairly representative of the total farming population. That said, Lambton UFO members tended to be slightly better off than their non-UFO counterparts. In Warwick and Plympton townships, the difference in the value of property owned by UFO members and non-members amounted to over three hundred dollars on average, while in Bosanquet township the difference approached sixty dollars.

The breakdown of Lanark UFO members by township is included in Appendices K to P (for townships in which ten or more UFO members were located). Overall, UFO members tended to be more or less

Table 16
Percentage of Members Who Sat on the Executive of Three Simcoe County UFO Clubs

	Total No. of Members	*No. of Those Serving on Exec.*	*% of Total Serving on EXEC.*
Rugby	43	14	33
Oro Station	61	24	39
Edenvale	47	22	47
Total	151	60	40

Source: Simcoe County Archives, Minute books of the Rugby UFO (1920–30), Oro Station UFO (1917–30), and Edenvale UFO (1920–34). Calculations by author.

representative of the general farm community. In Ramsay Township, where the most UFO members were located, the average member's farm was valued at $3,933.15, while a member of the general population held a farm valued, on average, at $3,664.21, a difference of $268.91.[19] In Drummond Township, however, the assessed farm holdings of UFO members tended to be worth less (on average, $412.50 less) than the non-members. Interestingly, in most cases UFO members tended to have smaller farm holdings with fewer cleared acres. Moreover, they held land and buildings worth considerably less, on average, than their counterparts in Lambton. But such figures may be misleading. The smaller farms may have been on better-quality land, and Lanark and Lambton counties may have employed different formulas for valuation. The only significant difference between UFO members and non-members relates to age. In townships where ages were recorded, UFO members tended to be younger than those sampled as non-members, which may also explain the fewer cleared acres on their farms.

It should be remembered that the UFO members found in both Lambton and Lanark were probably part of their local club executives. Were those farmers better off than non-executive members? It is likely that they were, judging from the three UFO clubs in Simcoe County – Edenvale, Rugby, and Oro Station – whose meeting minutes and membership lists are extant (Appendices H to J).[20]

As noted above, the comparative averages calculated for UFO members and non-members reveal that members tended to be younger than the overall farm population. A look at the Edenvale, Rugby, and Oro Station clubs suggests further that executive and members were younger than non-executive members. Granted, the sample in this case is small, but assuming that the results are not anomalous and that service on a club's executive indicates greater

Table 17
Property and Property Value Indicators Executive and Non-Executive Members in Three Simcoe County UFO Clubs

	Executive	*Non-Executive*
RUGBY UFO CLUB		
No. Identified	9	21
Avg. Age	36.38	41.33
Avg. Acres Owned	162.86	180.18
Avg. Acres Cleared	132.86	138.63
Avg. Non-Prod. Acres	30.00	41.55
Avg. Value, Land ($)	2,146.43	2,553.13
Avg. Value, Buildings ($)	896.43	1,012.50
Avg. Total Value ($)	3,042.86	3,565.63
ORO STATION UFO CLUB		
No. Identified	15	19
Avg. Age	37.07	42.53
Avg. Acres Owned	143.08	140.94
Avg. Acres Cleared	119.08	110.41
Avg. Non-Prod. Acres	24.00	30.53
Avg. Value, Land ($)	2,392.31	2,035.29
Avg. Value, Buildings ($)	1,069.23	820.59
Avg. Total Value ($)	3,461.54	2,855.88
EDENVALE UFO CLUB		
No. Identified	11	5
Avg. Age	40.82	53.40
Avg. Acres Owned	122.40	82.00
Avg. Acres Cleared	94.00	62.00
Avg. Non-Prod. Acres	28.40	20.00
Avg. Value, Land ($)	3,850.00	1,700.00
Avg. Value, Buildings ($)	810.00	733.33
Avg. Total Value ($)	4,660.00	2,433.33

Source: Simcoe County Archives, Minute books of the Rugby UFO, Oro Station UFO, Edenvale UFO, and various assessment rolls. Calculations by author.

commitment to and activity in the UFO than general membership does, one can infer that the UFO was a young person's movement. This feature of the movement invites some speculation. That youth was a defining characteristic could, for instance, reflect the idealism that pervaded the UFO. It could also represent an impatience – often associated with youth – for change (and a conviction that change cannot be achieved through traditional avenues), a fickleness of political allegiance, or a propensity for early disillusionment. As studies have indicated, young voters tend to be more volatile in their political allegiances, having fewer preconceived notions than older voters and being more flexible in their partisanship. Young voters also tend to be more apathetic than older ones.[21] As will be seen in the following chapters, all of these predilections found expression in the movement.

In addition to these differences in economic status and age between UFO members and non-members, there were also variations in religious affiliation. Assessors enumerated enough UFO members by religion in Lanark (a total of 118) and in the two Simcoe County townships under consideration (a total of seventy-one) to allow for a comparison to be made between adherents to the movement and the general population.[22]

In Lanark and the two Simcoe townships Presbyterians were significantly overrepresented in the movement, as were Congregationalists, albeit to a lesser extent. Exactly why this was the case is difficult to assess, but one plausible explanation lies in the state of Presbyterianism at the time. Richard Allen points out that Presbyterians made up a significant portion of Social Gospel adherents.[23] In addition, as John S. Moir notes, the denomination was embroiled during the early 1920s in a highly divisive debate regarding union with the Methodist church. He contends that those who tended to support union leaned towards the Social Gospel element of Presbyterianism to a greater extent than those who called for a separate Presbyterian church.[24] Moir's argument is supported by N. Keith Clifford, who further suggests that rural areas were more supportive of union than urban centres. Hence it seems logical, given the idealistic nature of Social Gospellers, that many of its adherents – rural, pro-union Presbyterians – found their way into an agrarian movement that was also characterized by idealism.[25]

Methodists played a significant role in the Social Gospel movement, yet they were slightly underrepresented in the UFO membership figures. Even so, as Table 18 shows, Presbyterians and Methodists combined accounted for sixty-six percent of UFO adherents in Lanark (the two denominations accounted for fifty-one percent of the general population) and eighty-three percent (as compared to fifty-three percent) in Simcoe. Obviously, the movement had a strongly evangelical composition.

Baptists were also underrepresented in these membership figures but not to the extent that Anglicans were. Why Anglicans accounted for such a small proportion of the UFO membership presented here is not known. Allen claims that, of all faiths, Anglicanism was among the least receptive to the Social Gospel movement. If this suggests that Anglicans were less receptive to social activism in general, it may explain, at least in part, their low interest in the UFO.

On the whole, though, it appears that the movement in Lanark and, to a lesser extent, Simcoe, was reasonably inclusive, especially

Table 18
UFO Membership – Breakdown by Religious Affiliation, Lanark County and Flos and Oro Townships

County Religion	*Number of UFO Members*	*% of UFO Members*	*% General Pop. in County*
LANARK COUNTY			
Presbyterian	65	55	34
Roman Catholic	18	15	17
Methodist	13	11	17
Anglican	13	11	23
Congregationalist	6	5	1
Holiness Movement	2	2	1
Baptist	1	1	4
FLOS AND ORO TOWNSHIPS			
Presbyterian	46	65	26
Methodist	13	18	27
Anglican	5	7	20
Congregationalist	4	6	1
Roman Catholic	2	3	17
Baptist	1	1	3

Sources: AO, Assessment Rolls; Simcoe County Archives, UFO minutes books; various newspapers. Calculations by author.

in terms of bringing Protestants and Catholics together in a common cause – no small accomplishment in a province where groups advancing religious and ethnic bigotry, such as the Orange Lodge, still wielded considerable political and social influence.[26]

The above suggests that shared religious affiliation was a significant feature in the composition of the UFO. It seems likely that membership prompted by or based on physical proximity to other members might also be a defining characteristic of the movement. The records of the three Simcoe County UFO clubs provide an opportunity to chart member pervasiveness. By matching UFO farmers to assessment roll entries, one obtains a notion of the proximity of members to one another. Figure 1 shows the location of identified members of the Rugby and Oro Station clubs in Oro Township. Farmers belonging to the Rugby Club are represented by an x and farmers belonging to the Oro Station Club are marked with an o.[27] As can be seen, known members lived fairly close together, and in some areas nearly every lot contained at least one person who at one time or another was a UFO member. The same pattern can be observed in the case of Flos Township's Edenvale UFO Club (figure 2), where members are represented by an x.

Figure 1
Rugby and Oro UFO Clubs Membership, Oro Township, Simcoe County

Conc.	VI	VII	VIII	IX	X	XI	XII	XIII	XIV
Lot									
6							x		x
7		o					x	x	
8		o						x	
9							x		
10						x	x x	x x	x
11						x	x		
12							x x	x	x
13								x	
14	o						x x		
15									
16		o							
17		o	o o						x
18		o	o						
19									
20			o						
21	o o	o oo							
22	o	o o		o					
23		o o		o					
24	o o	o o	o						
25	o								
26									

Note: Figure represents a portion of Oro Township only.

If Simcoe is taken to be a fairly typical county, it can be assumed that such clusters existed around centres in Lambton and Lanark as well. In Lanark, these clusters must have been fairly large because, as the next chapter will show, membership in UFO clubs in Carleton Place, Perth, and Smiths Falls at times exceeded one hundred. If the Simcoe UFO clubs were typical, and if clubs in Lambton and Lanark displayed similar characteristics, then the notion that the UFO was a pervasive organization is reinforced.

Figure 2
Edenvale UFO Club Membership, Flos Township, Simcoe County

Conc.	I	II	III	IV
Lot				
14				
15	x			
16	x			
17				
18	x			
19				
20	x			
21	x			
22	x x	x		
23	x	x		
24		x		
25				
26		x		
27		x		

Note: Figure represents a portion of Flos Township only.

It should be kept in mind, however, that the maps are not an exact picture of membership in these clubs because many members were not identified in the assessment rolls. Of the sixty-one members of the Oro Station UFO between 1917 and 1930, for example, only twenty-nine are noted on the map. If the other thirty-two members were plotted on the map, then the pervasiveness of the movement would be even more apparent. The same can be said of the other two clubs.[28]

Simply put, the pervasiveness of the UFO movement in rural Ontario is difficult to measure with any degree of accuracy, especially if it is based solely upon quantitative data. It is not known, for example, how often a UFO member may have mentioned to a non-member neighbour that his club was about to purchase a large order of binder twine, or how frequently, in such instances, the neighbour made a purchase through that member. In fact, some clubs actively encouraged non-members to make purchases through the local UFO. At best, one may suggest that the movement was a strong presence in rural communities. Yet it is plain from the examples of the three clubs in Simcoe County, as well as from other evidence to be cited below, that the number of member farmers identified above merely scratches the surface. Even the figure of sixty thousand members across the province at the UFO's peak cannot reflect the total number of members the movement had over the thirty years of its existence.

Although much can be learned from the preceding township figures, it should be kept in mind that the averages presented for each township are simply that – averages. Such figures tend to mask inequality, one of the more persistent characteristics of the countryside. That some farmers were exceedingly well off while others appear to have barely eked out a living is clear from many of the tables in the appendices. In Bosanquet Township, for example, the farmer Nicholas Sitter controlled, with his wife, 292 acres of land (237 of which were cleared) worth over ten thousand dollars, and buildings worth two thousand dollars. At the other extreme one finds farmers such as James Zavitz, who owned fifty acres of land (forty-five of which were cleared) and whose total assessable assets were valued at $1,850. In Ramsay Township Hiram McCreary's land and buildings were assessed at $7,800, while Dan McPhail's holdings were valued at $1,500. In Flos Township, Newman Giffen had farm holdings assessed at $10,600 and his brother, Henry, owned $8,300 worth of land and buildings. On the other hand, Zeeman Rupert's seventy-six-acre farm and buildings were valued, in total, at $2,700. And on and on. This sort of disparity was also found amongst the non-members who farmed in these townships.

If farmers with holdings of less than one hundred acres relied solely upon their farms for their income, then they may have found themselves in what Wylie terms a "poverty trap." A survey of dairy farms in Oxford County undertaken in 1917–18 showed that even in dairy enterprises, where acreage is not as important as it is for farms growing field crops for market, a holding of one hundred acres netted its owner only $1,296 for the year, after expenses. According to the survey, some farmers with less than forty-five acres actually lost money. As noted by A. Leitch, who oversaw the survey, "Unless a man has an enormously large farm, it is impossible to make over two thousand dollars (per annum) – and that only by working thirteen hours a day."[29] Other studies from around the same time confirmed Professor Leitch's conclusion – a farmer had to work extremely hard to make even a moderately sized farm profitable. Even then, the net returns could be quite small. A survey of Caledon Township in Peel County showed that the annual net return for township farmers averaged under one thousand dollars, less than the average annual income of an urban manual labourer. Surveys of Dundas, Middlesex, Dufferin, and Wellington County farmers from around the same time produced similar results.[30]

How, though, does this relate to the United Farmers of Ontario? Did the poverty trap have a bearing – perhaps even a marked bearing – on the character of the movement? The above clearly demonstrates that individuals with varying financial means joined the UFO. Rural society was stratified, and the composition of local clubs reflected this stratification. Despite the fact that rank-and-file UFO members might have been slightly less well off than non-members, economic status appears not to have been a determinant of membership. It also did not dictate the internal operations of the clubs. Admittedly, in many cases some of the more successful farmers occupied prominent positions in the local executives (or even at the central level). But at the same time, in actual day-to-day administration these people had, in theory, no more rights than other UFO members.

Of course, statistics cannot indicate why individuals joined the movement. In order to explain motivation, some speculation is necessary. Some farmers may have joined simply because the UFO offered them a professional group through which they could keep in touch with other farmers and remain current on agricultural issues. This motive probably did not account for many members, though, since at the time there were several professional associations that addressed the needs and concerns of particular farmers. Some undoubtedly became members in order to benefit from the savings realized through the collective purchase of consumer goods and farm supplies and to take advantage of the profits that might be made through collective selling. As noted earlier, however, members both ordered consumer goods and farm supplies for other farmers in their localities and marketed some of the produce their neighbours wished to sell. This seems to have been the case when the neighbours were relatives. Yet, as the lists of members in the appendices show, many families had more than one UFO member. Therefore, it is likely that there was more at work here than simply the desire to be a member of a professional organization or to maximize returns and lower prices for consumer goods and farm supplies.

Other factors may have prompted farmers to consider joining the UFO. Perhaps there was a certain amount of peer pressure, or perhaps being a member of a movement that was sweeping the province engendered a feeling of belonging. The social connection offered by the UFO might have been a strong motivational factor: members of UFO clubs regularly attended at dances, picnics, suppers, meetings, and so on, which brought them together so that the monotony and drudgery of farm life could be forgotten momentarily. The clubs also encouraged their members – particularly the younger ones – to write essays, perform musical numbers, stage theatrical productions, give speeches, and sit on committees of undertakings that were designed to benefit not only farmers but the larger community as well (such as beautification projects, the erection of war memorials, and famine and disaster relief).

Finally, discontent in all its various forms must have been a strong inducement for many to become UFO members. Some may have been outraged at the federal government's reneging on its promise to exempt farmers' sons from conscription. Others may have been disillusioned in the wake of the Great War, whose slaughter did not result in the democracy that had been promised by political leaders. Still others may have been disgusted by the reported graft and corruption among politicians and business leaders; uneasy about the movement of people away from rural townships to the province's cities and towns; or, perhaps, enraged at the impotence of farmers in determining how their province and country were governed. For all of these ills, farmers may have felt that the best remedy was participation in and promotion of a mass democratic movement such as the UFO.

To conclude, farmers in Lambton, Simcoe, and Lanark counties faced circumstances that were in some cases similar and in others distinct. In terms of social composition, all three counties reflected the norm throughout the older section of the province. That is, their populations were predominantly of British heritage and Canadian birth and split, in roughly equal proportions, between urban and rural residents; a few religious denominations dominated in each jurisdiction; and their rural areas were declining in total population. In terms of differences, Lanark farms tended to be larger than those in Lambton or Simcoe but were, on the whole, less productive and less valuable. Lambton farmers enjoyed not only better quality soil but also a greater choice of crops. Simcoe County farmers were only slightly worse off in this regard.

Some of this may have influenced the composition of certain UFO clubs. In several of the less well off townships of Lanark County, UFO members tended to be poorer than their colleagues. In other areas, where a certain degree of prosperity was enjoyed, there was a tendency for UFO members to be slightly better off than their counterparts along the concessions. However, overall (that is, considering both executive and rank-and-file members), it appears that a UFO farmer in all three counties was as close to an average farmer as one could be.

3 Ringing out the Narrowing Lust of Gold and Ringing in the Common Love of Good: The Rise of the UFO

(The "Big Interests") prate about restored prosperity and a full dinner pail, as if that was all you need.

From the Manifesto of the Smiths Falls Progressive Committee, 1921[1]

This chapter moves beyond statistics and demographies and examines the specific character of the UFO in Lambton, Lanark, and Simcoe counties during the time of the movement's dramatic rise. Since the period represented what Goodwyn would call "the movement educating," where members self-consciously developed an awareness and critique of the "prevailing forms of economic power and privilege" in society, I shall begin with an account of what farmers were saying about particular issues.[2] I then move to an examination of how, as they gained confidence and determination, farmers found concrete expression for their views, most notably in their spirited participation in the 1919 and 1921 election campaigns. Finally, I explore attempts by UFO members to develop alliances with urban workers as they strove to implement their ideas.

Throughout the period 1919–21 UFO members were increasingly suspicious of, and often outrightly hostile towards, not only politics as it was practised in specific instances but also established power in all of its various forms. The structures that, in the farmers' view, wielded inordinate influence – the old political parties, corporations, and the press – were often referred to collectively as the "Big Interests." UFO members blamed these Big Interests for much of the malaise that they saw around them and believed that their special privileges should be withdrawn and their abuses of power countered. Yet the movement, in the words of Simcoe's Sunnidale Corners UFO club members, "represented far more than a few disgruntled people wishing to teach the older parties a lesson," and more than a demand for occupational rights. It represented, more properly, "a great awakening."[3] UFO president R.H. Halbert threw out a challenge: "Canada was ... in a period of reconstruction, and were the farmers going to contribute anything to the moulding of public opinion that in reconstruction, the common people might get a square deal? ... Great issues were at stake today, and the next few years would determine the future of Canada. Had the farmers the courage to face the future with the inspiration of a great purpose or would they let past and present conditions go on?"[4] And farmers responded to that challenge. They hoped, as he did, "to see the day of cooperation among the common people," and to ensure the arrival of that day, they invested significant time and effort not only towards devising corrective measures for specific problems but also towards "saving" society as a whole.

At first, members' comments on contemporary issues were tentative, often mere reiterations of assertions made by the central leadership, however fierce the rhetoric employed. But given the frequency and intensity of their speeches, one wonders how often UFO leaders said what they thought members wanted to hear.[5] Perhaps the rank and file did have their own informed, strongly held views. Certainly over time – particularly during the 1921 campaign – as farmers became more sure of themselves, they began to articulate their own vision of a new democracy.

In effect, the 1919 and 1921 election campaigns marked a turning point for UFO members.[6] The Liberal and Conservative parties had failed "to play square with the common people"; as a result, "independent political action seemed to be the only course open to them."[7] The election campaigns showed farmers that there were others besides themselves who were dissatisfied with the *status quo*, groups with which they might establish alliances. Through the elections the UFO tried to play a more meaningful role in their own governance. The UFO rank and file laboured, with energy and enthusiasm, to create an alternative societal structure that would allow common people to participate in the formulation of public policy. In sum, Ontario farmers were experimenting with dissension in their critique of the *status quo* and they were giving expression to idealism in their attempts to develop and implement an alternative vision. The early 1920s were indeed an exciting time to be a UFO member.

THE GROWTH AND OPERATIONS OF LOCAL CLUBS

The UFO grew exponentially during the First World War. In February 1915 membership stood at roughly 2,000, but by 1916 the number had climbed to 5,000. In December 1917 there were an estimated 25,000 members in over one thousand clubs. Expansion continued at such a steady pace that in 1919 the UFO could boast of some 48,000 members. The movement reached its high-water mark after the 1919 provincial election; some one hundred new UFO branches were established in 1920 alone, and membership reached approximately 60,000. According to one contemporary estimate, the total UFO membership that year represented approximately one-third of the total number of farmers in the province.[8] It seems that the actual figure may have been closer to twenty percent, but, as noted earlier, there was turnover in membership, which means that more than 60,000 farmers were members at some point.[9] Moreover, the UFO had an impact even on non-members.

Clearly, this increased support for the UFO resulted to some extent from opportunism, given the 1919 election victory. But new members were more likely to be motivated by the sense of optimism that had been generated by the movement, an optimism that manifested itself in increased discussion and debate of public policy issues with the goal of enhancing popular participation in the political system.

Provincial figures do not reveal much, however, about the movement's growth at the local level. Exact membership figures for most clubs in Lambton, Simcoe, and Lanark counties are difficult to ascertain, but in Lambton there were four UFO clubs in the Forest area alone by July 1915, and within two years membership in and around that town numbered some three hundred. By late 1919 there were over seventy-five UFO clubs in the county.[10] If the average membership of each Lambton club was thirty, then the county membership was roughly 2,250. Although the UFO had not been a significant force in Simcoe County prior to 1917, new clubs emerged soon after. The Guthrie Farmers' Club affiliated with the UFO in 1918, and new clubs were formed in Loretta, Kirkville, and McMurchy's Settlement that same year.[11] In 1919 it was estimated that there were some six hundred UFO members in the provincial riding of South Simcoe.[12] In Lanark, the Carleton Place UFO Club had over one hundred members by early 1919 and was approaching two hundred by June of that year. The clubs in and around Pakenham at the time had over two hundred members, and Farmers' Clubs throughout the county continued to affiliate with the UFO.[13] One UFO member estimated that there were nine hundred members in Lanark in 1919. By 1921, the county sported eighteen clubs, many of which had memberships of over one hundred.[14]

Membership figures tell only part of the story regarding local connection with and involvement in the UFO. Picnics and other social events could, and often did, attract thousands of people.[15] These events brought farmers and their families together, thus relieving feelings of isolation. They also allowed UFO members and non-members alike to share ideas about politics, economics, and other issues of the day.

How did the clubs operate? It appears that local members had the autonomy to structure their clubs and meetings as they wished, within some broad boundaries. Extant minute books provide few clues about matters even as basic as the duration of meetings. It seems, however, that at least a part of each meeting was devoted to cooperative matters. After they had been dealt with, meetings might proceed in a variety of ways. There might be guest speaker – perhaps a central UFO personality or the local agricultural representative; a local issue might be discussed; a committee might report on its work; electoral strategy might be formulated. In effect, meetings were what local members wanted them to be.

Considering that the UFO contained so many young people, it is not surprising that clubs involved themselves in many initiatives related to self-improvement, sport, and entertainment. Some clubs held regular debates, so that farmers could discuss contemporary issues and develop elocution skills. As one member put it, "If we could all speak in public, it would make a great difference in the way politicians and town folk think of us."[16] In some localities, sporting clubs or even sports leagues were organized. Other clubs, such as the Kinnaird UFO in Lambton, formed literary societies that staged readings, recitations, and concerts in front of large and appreciative audiences.[17] Not all clubs were equally ambitious, but most featured annual musical or dramatic performances.

In their role as service organizations, UFO clubs often acted on behalf of the larger community. In Simcoe, for instance, the Rugby Club was instrumental in erecting a monument to those from the village who had served in the Great War.[18] Similarly, members of the Lanark County Cedar Hill UFO Club explored the possibility of obtaining electricity for the locality, entering into negotiations on the matter with Carleton Place mayor G. Arthur Burgess.[19]

Of course, in addition to providing services, the clubs were politically active. A countywide legislative committee in Lambton, for example, was designed to handle farmers' complaints with public policy by taking "the cases up directly with Ottawa."[20] The annual picnics and special winter-month suppers that virtually all UFO clubs staged had political overtones, even in non-election years, and even before the UFO explicitly entered politics.[21]

DEVELOPMENT OF SOCIAL CRITIQUE

During the period of the "movement educating," to repeat Goodwyn's phrase, the UFO rank and file attained "a heretofore culturally unsanctioned level of social analysis."[22] During the First World War, farmers in Lambton, Simcoe, and Lanark grew increasingly uncomfortable with the restrictions placed upon them by the state – the prime issue being the revocation of conscription exemptions for farmers' sons and farm labourers – and began to grumble against established power. Movement leaders such as E.C. Drury insisted that it was vital for all UFO clubs to "thoroughly study and discuss the affairs of the country [so that] they may be able ... to discharge their great duty to the cleaning up of the affairs of Canada."[23] Through the efforts of such leaders to expose UFO members to a host of new ideas, farmers began to formulate their own critique. At first they presented their views – whether on the biases of the urban press or on the country's corrupt political system – hesitantly and cautiously. Gradually, however, they became more forceful; as one member noted, the issue had never been one of intelligence or resolve but of confidence.[24] Indeed, their confidence grew to such an extent that at one Simcoe County meeting UFO president R.H. Halbert was denounced as being a detriment to the movement because he tended to stifle independent thought.

By the 1920s farmers were spontaneously formulating and expressing their views, so much so that one UFO MPP, W.I. Johnson, admired the way formers were "studying the affairs of the world today more than ever before [and] were doing their own thinking."[25] The main themes of the UFO's critique of society are as follows.

Rural/Urban Tensions

The tensions between rural and urban society have been described in several studies of the UFO. Briefly, farmers were concerned that cities were draining rural areas of their sons and their farm labourers. Equally important, many farmers believed that urban centres, whose political influence was disproportionate to their share of the population, were gaining even more power over public policy formulation. UFO members pointed to these inequities and argued that cities – with their slums, abject poverty, and loose morals – were precisely where many social problems had their origins.

Although these views were fuelled to some extent by the *Sun*[26] and by UFO leaders, their distrust of urban society derived largely from the experience of farmers themselves. One Lanark County farmer was bemused when he and his "hayseed" colleagues heard of the proposal, advanced by a group of urban businessmen, that farmers should be compelled to raise all the livestock they possessed to maturity. "Alas," he wrote, "it takes more than wind to feed these animals and fowl."[27] Farmers often received bad press in newspapers.

One Smiths Falls columnist claimed, for instance, that UFO members were convinced that city people "were lying awake at night plotting against their peace and happiness."[28]

Rural/urban tensions emerged clearly in the "Patriotism and Production" campaign, a federal/provincial initiative designed to encourage farmers to increase productivity in order to satisfy wartime demand. Some farmers complained that while urbanites called for greater production, they did little or nothing to assist the farmers in attaining that goal.[29] Many others protested silently by treating the campaign with indifference.[30] Farmers argued that they were doing their best to increase production, and the notion that they needed to be told by urban-based bureaucrats to heighten their efforts undoubtedly created resentment not only towards those implementing the program but also towards urbanites in general.

Rural hostility towards urban society was in large measure a matter of self-defence. Cities, with all their conveniences and employment opportunities, presented rural youth with an attractive alternative to farm life. To counter their allure, it was believed that the benefits of rural life needed to be espoused, and there was no better way of accomplishing this than setting up a negative comparison. Given the prevailing view that there was something about rural life that made it morally superior to city life, mounting such a "campaign" was not difficult.[31]

Pitting rural and urban forces against one another enabled farmers to conceive of their plight as something that originated elsewhere. The city, with its concentrations of wealth, power, and political sway, was surely responsible for at least some of the problems farmers faced. But while casting urban society in an unfavourable light did much to bolster the farmers' self-confidence, it was also problematic. In the case of UFO members, as will be seen, ascribing blame to cities and urbanites increased the difficulty of forming an alliance with labour.

Believing that urbanites saw them as uncultured and uneducated rubes, many farmers cast jaded eyes upon relatively innocuous initiatives. In 1918, for instance, the Ontario Department of Agriculture sent farmers calendars that contained farming tips, dates of agricultural meetings, and so on. Many recipients claimed that the calendars were an insult to farmers' intelligence and that advice such as "Do not kick over the lantern" and "This is a good month to oil the harness" was pure condescension. Farmers wondered why other classes had not been sent similar publications. That the calendars cost the province some $18,000 to produce and distribute only added to the farmers' outrage and lent credence to their argument that the exercise was nothing but a cynical vote-buying effort by the government of William Hearst.[32]

Conscription

Clear evidence – at least as far as farmers were concerned – of urban ignorance of rural conditions emerged during the debate about whether or not Canada should conscript men for service in the Great War. In the view of farmers, distant policy makers who had no knowledge of what they endured were regularly making decisions that affected their ability to farm. To make matters worse, while telling farmers to increase production, the government was proposing to undercut their ability to do so by taking away their sons and their hired help. As it became increasingly obvious that some form of conscription would be introduced, the farmers' sentiments boiled over.

Although they risked being labelled unpatriotic, a significant proportion of Ontario farmers responded angrily to conscription. For many, it was their first experience of outright dissension. They viewed the actions of the government as arbitrary, unfair, ignorant, and undemocratic, and they let their opinions be known. The situation ultimately provided some important insights and lessons for UFO members, many of whom began to question other "truths" that were central to the society in which they found themselves.

Even before conscription was extended to hired hands and farmers' sons, farmers were grumbling about the government's expectations. By late 1917, for example, it had become common practice for military tribunals to cite the ratio of a "man-and-a-half per hundred acres" as a sufficient measure of manpower required to raise and harvest crops. "A Farmer Who is Awake" in Simcoe County asked why farmers had not been consulted when this ratio had been determined. Comparing farms to factories, he inquired whether any other manufacturers would "allow themselves to be treated with such contempt, and not even demand a voice in the councils that deal with their vital interests, their rights, their very self-respect?"[33]

By late 1917 many farmers had begun to doubt the sincerity of the Union government's promise not to conscript farmers' sons.[34] Local feeling was vehemently negative. In November 1917 the Perth Farmers' Club held a meeting on the subject, which was attended by some five hundred people. A petition protesting "the manner in which the local tribunals are interpreting the Military Service Act" was drafted at this gathering. During a sitting of a tribunal at Perth, the petition noted, some one hundred applications for exemption were considered. Twenty of the men who had applied were ordered for immediate service, and the remainder were granted exemptions for periods of only two to six months. Approximately seventy-five percent of the applicants were farmers' sons, and local farmers were outraged that their applications were being treated in such a manner.[35]

A well-attended meeting on exemptions was also held at Tottenham, in Simcoe. The unanimously passed resolution that was the outcome of this meeting was sent to the minister of Agriculture. After noting that much remained to be done to complete that year's harvest, and that other work on farms had been neglected in the attempt to produce more food, the resolution highlighted the fact that many farmers' sons and farm labourers had been denied exemption: "It is our petition that all these boys should be allowed exemptions from military service when their appeals are reached ... We declare our loyalty to the purpose of winning the war ... at the same time urging that it is only consistent that the producers of farm products should not only be allowed, but compelled to remain on the farms."[36] Given that farmers were trying to fulfil their wartime obligations, they deserved fairer treatment.

Clearly, conscription was a sensitive issue for farmers, and it became even more sensitive as the perception grew in rural communities that the government was reluctant to ensure that all Canadians paid their fair share towards the war effort. At the 1916 UFO provincial convention, members went on record as opposing "any proposal looking to the conscription of men for battle while leaving ... plutocrats fattening on special privileges ... in undisturbed possession of their riches."[37] Local members added their own unique touches to this sentiment. "J.K." of Lambton, for one, believed that people with surplus revenue should lend it to the community for a period of five to ten years, interest free. He felt that if the present government could show the courage to enact such a measure, then there might not be so much grumbling over conscription. He believed, however, that no jurisdiction in Canada was capable of taking this step because, although Canada had sent troops overseas to fight for democracy, the government was "practising tyranny at home."[38]

When it finally came, the revocation of Prime Minister Borden's promise not to conscript farmers' sons was met with vigorous protest by agrarians. Protest meetings were held throughout Ontario. In Simcoe, for instance, some 350 farmers attended a rally in Oro Township to protest the Military Service Act. According to press accounts, only one speaker supported the new regulations, and he was given rough treatment by the rest of the audience. A resolution, passed unanimously, accused Borden of violating his campaign promises. It went on to assert that Oro farmers were neither motivated by selfish interests nor disloyal. Instead, they aimed "to avert a wholesale tragedy" that would result if production was allowed to diminish.[39] Consequently, Oro farmers demanded that the government "honor its pledge to exempt all ... farmers" from military service.[40] One newspaper noted wryly that many who attended the meeting were of draft age, and it questioned the loyalty of the participants. A gathering held in Uhthoff a few days later – labelled an "anti-draft meeting" by the local press – attracted well over one hundred farmers, mostly from Orillia Township. A resolution similar to that developed at the Oro Station rally was approved, with only two dissenters.[41]

Individuals also expressed their urgent and heartfelt concerns over conscription. James Sheehan, reeve of Adjala Township in Simcoe County and a staunch UFO supporter, chided the government for its hypocrisy and declared that he had heard of a farmer from Adjala who had approached an officer at Niagara Camp to ask how he was supposed to harvest his crops: "The officer is quoted as putting him off with 'it is men we want, to [Hell] with production.' This, I imagine, is what the Kaiser would advise ... No doubt the Kaiser would be glad to see our cattle destroying our crops." The controversy prompted Sheehan to remember the 1917 federal election campaign: "Where are the men today who last fall were going up and down the back concessions, shaking hands, patting us on the shoulder, and whispering into our ears promises – promises they have since broken? Where are they today?"[42]

All of this, of course, raises the question whether farmers were motivated purely by self-interest or whether an aversion to militarism played a role in their response. Agrarians tend to be given short shrift in most accounts of pacifism in Canada, but there is little doubt that some UFO members, with their idealist bent, opposed war as a matter of principle.[43] Calling himself "Conscientious Objector," a farmer from Lambton County went so far as to ask the *Sun*'s editor if he knew of any religious groups that opposed the war. He had been a lifelong member of the Methodist church, but "in consideration of the stand it has taken regarding war and politics I feel as a matter of conscience that I must sever my connection with it." Another correspondent to the *Sun*, "Atom," noted that earlier cultures had made sacrifices to war gods: "So doth the modern dweller of today with the same easiness of belief, cast their offspring into that hell hole [war], to satisfy their gods of profits and empires."[44]

Note that these and several other letters expressing similar sentiments were written during the war and not after, when it was much easier and less dangerous to adopt a pacifist stance. However much one might argue that this was nothing more than farmers' trying to justify a position against conscription, it must be stressed that such views were highly unpopular in English-speaking Canada at the time, and the dissidents who voiced them experienced considerable backlash. Nor did the pacifist stream in the movement dissipate with the end of the war. UFO leader R.H. Halbert found receptive audiences when, in 1920, he condemned compulsory military training, arguing that militarism was the primary cause of the Great War.[45]

The Tariff

Another manifestation of rural/urban differences – and another controversy that UFO members entered into – was the debate over Canada's protective tariff. In many works on early-twentieth-century agrarian protest movements the tariff occupies centre stage.[46] By these accounts, concern over the tariff bordered on an obsession for many farm leaders (including UFO officials) and, by implication, occupied a central position in the thoughts of the rank and file. Certainly, the movement's leaders spoke and wrote widely on the tariff, and their assertions reached local members. Some UFO members argued that the tariff was a red herring offered up by the old political parties as either the saviour or curse of the country. In other words, it was another example of the Big Interests controlling the political agenda. Rank-and-file members saw the tariff controversy as part of a larger malaise that went beyond mere trade barriers.

Some of the most vocal tariff critics at the local level were not farmers, although they were sympathetic to the UFO. H.J. Pettypiece had nothing but contempt for the government for placing high tariffs and taxes on agricultural implements during the war, when the machinery was most needed. From April 1916 to January 1917, he noted, Canadian farmers paid nearly $1,250,000 in duties and war taxes on implements: "Increased food production is an absolute necessity in order to win the war, but until our Parliaments free themselves from the grip of the combines which now control legislation, no great increase need be expected. The Kaisers we have created in this country are as detrimental to our freedom as is the German Kaiser. Parliamentary Government in this country is a howling farce."[47]

Editorials such as this could not help but have some effect on the UFO members who read them. Regarding the fight against high tariffs, one Simcoe County farmer asked, "Why cannot the consuming public in the cities see that the farmer is not only fighting his own battles, but theirs as well; in saving ourselves we save the customer also."[48] In fact, the effort to which some farmers went to point out distortions and propaganda regarding the tariff was, at times, quite amazing. D.A. Taylor, a farmer from Lambton, was angry with the *Sun* for printing advertisements of the Reconstruction Association, which was a front, he argued, for protectionist forces. Although the association argued for protection, Taylor discovered that "a great many of these manufacturers have absolute Free Trade in their raw material ... If the *Sun* would publish ... pages 36 and 37 of Schedule A of the Customs Tariff opposite the Association's ad ... and follow it up with a facsimile of Schedule B the next time, it would certainly top their ace ... and give the farmers of Ontario the best object lesson they could have."[49] It is not known how Taylor reached his conclusions. Perhaps he had been shown the schedules at a UFO meeting, or maybe they had appeared elsewhere; conceivably Taylor had come upon them himself. At any rate, he retained the information and imparted it to his colleagues as yet another example of the hypocritical tactics to which pro-tariff forces resorted.

It appears that UFO leaders were more interested in the tariff than rank-and-file members were. There are several reasons for this. First, since many of the UFO élite were prosperous farmers, they saw the connection between the tariff, markets, and farm machinery prices more readily than average agrarians did. In addition, they had better access to data concerning the effects of high duties. However, even though their views on the tariff may not have harmonized, the UFO leaders still had a marked impact on local members. By writing widely on the topic, they gave the rank and file yet another reason for distrusting those who had implemented and controlled the tariff. Although the issue was not much debated at club meetings, it is reasonable to state that, for most UFO members, the tariff was yet another example of an indifferent or biased government that did the bidding of the Big Interests. It was proof for UFO members that the government, though allegedly for "the People," had actually been perverted by corrupt forces.

The Big Interests

At the local level, one finds numerous references to the Big Interests, an abstraction that corrupted political morals, exploited farmers and workers alike, and served as the motive power behind a fiscal system that rewarded only a small privileged sector of society. As with other issues, UFO members were exposed to the angry rhetoric of their leaders and other progressive commentators who enabled farmers to think about their situation and, in the process, to find their own examples of the negative influence of the Big Interests.

One central concern of many UFO members was corruption in both business and government, and for many these issues were intertwined. In one notable case, Joseph Flavelle's meat-packing firm, the William Davies Company, was accused of profiteering for allegedly having realized a five-cent profit on each pound of bacon it sold. H.M. Gadsby, a prominent *Saturday Night* columnist, was a persistent and often vicious critic of Flavelle. His columns were reprinted in several rural weeklies, so many UFO members were exposed to the machinations of Flavelle and his business associates. And, of course, these machinations added grist to the farmers' mills.[50]

The UFO leadership certainly encouraged farmers in their suspicion of the capitalist class. Speaking at a UFO picnic in Simcoe in 1919, E.C. Drury noted that the Dominion Textile Company (which, he alleged, practically controlled the Canadian cotton trade) paid an average annual dividend of twenty-three percent from 1916 to 1918. Even so, it was one of "the infant industries … howling for more protection." Speaking at the same picnic, J.J. Morrison claimed that during the war the wealthy in Canada were allowed to "tie up their money in war securities exempted of taxation" to the tune of $1.4 billion. Soldiers, who received $1.10 per day and who came home physical and emotional wrecks, were now expected to help pay off the war debt: "It is your fault. You allowed a pack of rascals to rob this country in the name of patriotism." At a Lambton UFO picnic a short time earlier, R.W.E. Burnaby noted that Dominion Textile had realized a three hundred percent profit the previous year, which fuelled the high cost of living for farmers and the general public.[51]

Other commentators, such as H.J. Pettypiece, augmented the leaders' accusations with some denunciations of their own. Responding to Lord Shaughnessy's comments that farmers were in a rut, Pettypiece agreed: "If we had not been for years in ruts we would not have given a thousand million dollars to railway exploiters and there would not be so many 'Lords' and 'Knights' and other tinhorn title bearers in this country … if we were not in ruts we would not allow the railway and other combines to control most of our legislation."[52]

It was apparent to the average farmer that many politicians were in the pockets of industrial concerns, and farmers often made their views on this public. During the 1921 federal election, for instance, North Simcoe UFO members ran a "UFO Column" in a local weekly in which they launched several attacks on the local Conservative MP, Col John A. Currie. It was noted, for example, that Currie shed "crocodile tears" when the Massey-Harris agricultural implement factory closed down. He had blamed farmers, with their supposed propensity for purchasing American products, for the company's failure.

UFO members claimed in turn that the president and general manager of the company ("two more patriots") drove American cars, proving that they did not practise what they preached. The attacks on Currie included an allegation that, although he professed to be anti-combine, he had been involved in a scheme to corner the wire-fencing market.[53]

How, in the opinion of UFO members, had the Big Interests accomplished the "industrial piracy"[54] by which all average Canadians were exploited? For many farmers, two separate but related forces accounted for the current state of affairs: the press, and a government willing to be bought by the highest bidder through political parties. In fact, it did not take farmers long to realize that many politicians from the old parties, as well as members of the partisan press, were actually part of the Big Interests that they served so loyally.

The Press

The leaders of the UFO helped to inculcate many farmers with the idea that the press could not be taken at its word. Again, their rhetoric struck a responsive chord. The very press that sanctimoniously congratulated itself for shaping public opinion was perceived by farmers as a vast propaganda machine that narrowed the boundaries of debate and frequently resorted to falsehoods and fabrications to accomplish its ends. John Kennedy, vice-president of the Grain Growers Grain Company (GGGC), told a Perth audience in 1917 that "with the exception of some farm papers and local town papers [Canadian newspapers] are invariably subsidized ... We usually find that each paper is controlled by a political party; each one is abusing the other ... Then we have other presses that are supported by large capitalistic corporations."[55] With such connections, no newspaper could be trusted to provide anything resembling the truth.

Farmers were presented with countless incidents showing the press to be in the back pocket of industrialists and political parties. H.J. Pettypiece, a newspaper man himself,[56] argued that urban dailies were under the control of manufacturers, and he provided several examples to bolster this contention.[57] A.A. Powers of the Sun Publishing Company (the farmer-owned enterprise that published the *Sun*) liked to point out distortions in the urban press.[58] During the 1919 provincial election campaign, UFO candidates often referred to control of the media by the "Big Interests."[59] Like Pettypiece, they stressed the need to present the "real facts" to the people so that they could make informed decisions regarding public policy issues. By 1921 distrust of and contempt for the press had intensified to the point where some Progressive candidates promised that, if elected, they would enact legislation requiring newspapers and periodicals to state publicly who owned them.[60]

The leadership's critique of the media was taken up by UFO members[61] as they began providing their own examples of the questionable credibility of the press. E.A. Elsom of Lambton showed how urban papers took unfair shots at rural residents. Others, such as "Farmer" in Simcoe, noted that the urban press often advised farmers to buy needless "labour saving" implements. Some farmers ruined themselves financially making such purchases. This, the correspondent claimed, was one cause of rural depopulation.[62]

The critique of the press by rank-and-file members intensified after the 1919 provincial election and found its clearest expression during the 1921 federal campaign. It appears that farmers were motivated to adopt this stance in part because many newspapers took an anti-UFO position and were eager to report scandals or rifts within the movement.[63] Whatever the reason, UFO member opinion of the press was quite low by 1921. After a North Simcoe nomination convention that year, Collingwood's Mayor Holden complained to the *Orillia Packet* that many speakers at the meeting heaped abuse upon abuse on the press. Of the two journalists who had attended the meeting, one left early in disgust. The one who remained asked for evidence to back up the invective being hurled at newspapers and was shouted down for his troubles. Clearly, the subject was a sensitive one for local farmers.[64]

In the *Collingwood Bulletin*'s "UFO Column," a space that had probably been purchased by the UFO, it was noted during the 1921 election campaign that the *Collingwood Enterprise* had recently vilified the GGGC with factual distortions. After refuting the charges, the column's author accounted thus for the misrepresentations: "It is quite probable that the *Enterprise* hoped ... to extract a little sympathy from the electors of North Simcoe for the candidate supporting the Big Interest Government. Their efforts are in vain. The people are wise to this political game."[65] No longer would the media be taken at face value.

The Old Political Parties

The last third of the Big Interest equation was a political party system that not only turned a blind eye to corruption but actually participated in it. During the 1919 provincial campaign West Simcoe UFO candidate Richard Baker pointed out that, although both old parties officially condemned participation by government officials in elections, "yet, right here in West Simcoe we have a paid official of the Hearst Government acting as President of the Liberal-Conservative Association."[66] D.A. Taylor of Lambton summed up what the 1919 provincial contest meant to him: "The coming election has to be fought against the money power and there are those who say the money power will win every time. It also has to be fought against a corrupt and lying press ... Will the people stand true to their own interests in the face of it all?"[67]

The connection between the Big Interests and political parties was featured prominently during the 1921 federal campaign. Leaders such as T.A. Crerar pointed out that the Big Interests were "bearing the expenses of this government's campaign" and that, in return, the government allowed corporate mergers that further exploited Canadians. East Simcoe UFO candidate Thomas Swindle informed audiences that railroads were constructed with massive public subsidies to the Big Interests, who amassed enormous wealth from them and then sold them to the government when they were no longer profitable.[68] UFO supporter James T. Gunn of Simcoe argued that the Conservative party's *raison d'être* was to benefit the Big Interests and the ruling classes.[69]

For many converts to the UFO, the old parties represented outmoded and dangerous forms of political behaviour. James Martin, originally nominated to contest Centre Simcoe for the UFO in 1919, later resigned. Among his reasons for doing so was that if the Liberals did not nominate a candidate for the riding, he would be labelled a Grit, due to his past affiliation with that party. In 1919 R.W.E. Burnaby told a Lambton audience that anyone believing that the Liberals or Conservatives would change the existing state of affairs was delusional. Simcoe's Compton Jeffs spoke of the massive Tory effort to defeat farmer candidates in Manitoulin and North Ontario.[70] North Lanark UFO candidate Hiram McCreary admitted to having voted both Grit and Tory in the past but vowed that he had learned from his mistakes. He warned his supporters that both old parties were well organized for the election, and that they would resort to tricks or deception to attain their ends. In short, the old parties were institutions bent on forestalling any political, social, or economic advancement for the common people.[71]

Prohibition

"It would be difficult," Richard Allen contends, "to exaggerate the significance of the prohibition movement for the Social Gospel."[72] Given that the UFO was drawn heavily from the evangelical denominations, it is logical that prohibition or, at the very least, regulation of the manufacture and sale of intoxicants was a concern of many farmers.

Virtually every UFO candidate in the 1919 election campaign stated his views on the subject, and those who were ambiguous were given a rough time by rank-and-file members. East Simcoe candidate J.B. Johnston, for instance, was heckled by his own supporters when he refused to state his position on temperance.[73] Generally, being a strong advocate for temperance legislation was seen as a positive attribute by many UFO candidates and members. Even Liberal and Conservative candidates saw political advantage in noting that they supported the Ontario Temperance Act.[74] And if interest in temperance legislation was strong in 1919, it had intensified by the 1921 federal election. Lanark Progressive candidate R.M. Anderson was not alone in advocating the total prohibition of the manufacture, importation, and sale of all intoxicants.[75]

Whether it was the UFO's stand on temperance or its policies in general that attracted religious figures to the movement, the fact remains that several ministers appeared, apparently enthusiastically, at local meetings and at other UFO-sponsored events. The Reverend Mr Greig of Balderson supported the values advanced by the UFO, and as early as 1918 he reminded Perth area farmers that "the making of better citizens should stand higher in the aims of the order than any mere pecuniary gain." In Lanark in 1919, a UFO meeting in Maberly featured a Reverend Mr Clark, an Anglican priest, and Reverend Mr Smith, a local Methodist preacher. In Simcoe, the Reverend N. Campbell gave words of encouragement to Oro farmers shortly after the 1919 election.[76] In some ridings during the 1921 federal election, clergymen spoke publicly in favour of Progressive candidates because of their commitment to ending political corruption and, more importantly, because of their stand on prohibition. Lanark's R.M. Anderson enjoyed the support of Middleville's Presbyterian minister, the Reverend J.B. Townend, who "looked upon the farmer movement with friendly eyes," and Middleville's Congregational minister, the Reverend Mr McColl, who was fond of Anderson's stand on prohibition and gambling.[77]

Why did farmers feel so strongly about the need to regulate or ban the sale of intoxicants? Part of their reasoning may relate to their suspicion of large-scale industries, of which breweries and distillers formed a sizeable sector. Farmers also tended to see intoxicants and taverns as yet another example of urban society taking humanity down the wrong path. Finally, thanks to their religious background, most farmers saw alcohol as an evil that tended to erode Christian values. This third possibility bears closer examination. There was indeed a strong moral overtone to UFO member views on alcohol. Moreover, it is indisputable that the movement had an evangelical flavour. Lambton County organizer J.J. Wilson asserted in 1920 that his heart and soul were in the UFO; "he considered it the only source of salvation to better the conditions of this country and her people."[78] In taking a prohibitionist stance, farmers were clearly not acting out of economic or political self-interest. From an economic perspective, prohibition of the sale of alcohol meant that distillers and brewers would not be buying the grain that farmers produced. Politically, a strong stance on prohibition did not make much sense since it meant that many labourites in urban areas – who tended to be anti-temperance – would find it hard to form a lasting political alliance with farmers. Religious and moral convictions must, then, have taken precedence in the formulation of their views.

THE RANK AND FILE

One of the central points of this study is that local UFO members and UFO clubs often formulated ideas of their own that were striking for the sophistication of their analysis and for the thoroughness and clarity of their proposed solutions. Ideas that augmented official policy came from a number of sources. Some were the conceptions of a single person; others seem to have arisen spontaneously within local clubs; still others were attempts to refine policy or to make it work in a specific locale.

Several rank-and-file UFO members can be characterized as original thinkers, those who observed contemporary political, social, and economic behaviour and found it lacking. Certainly, some of the solutions they proposed were anything but conventional.

F.E. Webster of Simcoe County marched to his own drummer.[79] He frequently contributed letters to the *Sun*, and his insights presented a unique UFO alternative to the *status quo*. In 1918 he predicted that returned soldiers would shoulder much of the war debt rather than those who had reaped large profits from the conflict. As a remedy, Webster advocated a universal tax on land values. Noting that United Farmers in the West supported this measure, as did the "single taxers, the socialists, the anarchists and the labor unions," he suggested that such groups follow the lead of the prohibitionists with their Committee of One Hundred and form a Committee of One Thousand to press the government to implement the tax.[80]

To support his position, Webster often cited writers with whom he agreed, including the American anarchist W.C. Owen. One such passage read: "Anarchists do not propose to invade the individual rights of others, but they propose to resist ... all invasion by others. To order your life as a responsible individual without invading the lives of others is freedom; to invade and attempt to rule the lives of others is to constitute yourself an enslaver; to submit to invasion and rule imposed on you against your own will and judgement is to write yourself down a slave."[81]

A vocal critic, Webster addressed a variety of topics. In addition to those cited above, he challenged the myth that farmers were profiting from the war and he praised Wilsonian reforms in the US, noting that Canada had no equivalent figure: "Privilege goes on its merry old way ... Freedom! How much of it do we enjoy? How much freedom have we won since this great war began? Are we fighting at home as vigorously as the veterans in France to rid our land of autocracy?"[82] It is likely that he spoke for many who were less articulate than himself.

Other farmers, although perhaps not as perceptive and original in their thinking as Webster, also developed their own ideas. D.W. Lennox, a Lambton UFO member, was outraged that it cost Ontarians five thousand dollars just to heat the lieutenant governor's official residence; this was surely "one of the biggest humbugs ever worked off in a democratic country." By way of a solution, he proposed that the official residence be sold to some millionaire "with money to burn," and that a much less expensive residence be purchased for the lieutenant governor, who would then have to pay to heat it himself.[83] His suggestion was not revolutionary, but it shows that individual members were thinking about issues and devising remedies. It is also probable that Lennox's self-confidence was bolstered by the exercise, and that many farmers like him had similar experiences. In this way, voting for an alternative to the old parties became all the easier.

Another example is W.I. Johnson, candidate for South Lanark in 1919. Johnson told audiences how he reconceptualized his notion of democracy while serving overseas during the war. In Europe he had the opportunity to meet all sorts of people and to discuss politics. Through this process he realized that Canada was not as democratic as it could be. He came to see that the government represented the profiteer and the Big Interests, not the people. If elected he promised to "go to Toronto and stand on the floor of the house and thrash out questions" so as to ensure that the people of South Lanark were represented in the Legislature. Later in the campaign, Johnson revealed his personal preferences to the voters. He favoured public ownership of utilities (in order to destroy combines and profiteers), government control of natural resources, direct democracy, and a patronage-free civil service.[84]

Other farmers broke free entirely from conventional rural thinking and, to some extent, from farming interests. Alfred Wilkes of Lambton cancelled his subscription to the *Sun* because the paper opposed the legalization of margarine: "I thought your paper was strictly non-partizan, but your attitude ... proves you are as strongly one-sided as either Grit or Tory." Why Wilkes supported the legalization of margarine is not known. He probably knew, however, that it would adversely affect dairy farmers; his opinion thus points to a true independence of mind.[85] D.A. Taylor of Lambton differed from leaders and local members alike when he advocated political action, but with the UFO as a coordinating body, not a party *per se*. He was convinced that, if the UFO began acting like a political party, even if it was not an official one, then the opposition would smash it. Instead, Taylor proposed nominating farmer candidates who, instead of pledging to adhere to a platform, would merely act in the interests of agrarians. At least one club attempted to act on this principle: West Lambton's candidate in the 1919 provincial contest, J.M. Webster, was officially listed as an independent "in the interest of the UFO members of the district," a distinction not found in accounts of other ridings.[86]

ACTING ON THEIR SOCIAL CRITIQUE

Identifying societal problems is one thing; acting to remedy the areas of concern is quite another. In the case of the UFO, the initial response was to fight the Big Interests on the political front. By 1919, as W.C. Good pointed out, farmers had realized that it was foolish "to nullify their political influence by splitting their votes between two political parties who were practically identical in their general character." As the East Lambton UFO Political Association put it, "the time has arrived when the organized farmers should take steps to place a candidate in the field."[87]

If more agrarians were elected to legislatures, then government policy would more accurately reflect their views, or so the argument went. In addition, if other occupational groups followed their lead, the polity would accurately represent the collective will of "the People." Thus, in order to appreciate the full extent of agrarian discontent and the alternatives that UFO members advanced, it is useful to examine the 1919 provincial and 1921 federal elections. Elections provide valuable insights into the thinking of individual members and the autonomous operation of local clubs. In every riding in the three counties under consideration, UFO/Progressive candidates were nominated for the 1919 provincial and 1921 federal contests. For many members, it was only logical that farmers take this step. As one Simcoe farmer wrote: "Many farmers think the UFO should keep out of politics. But why? Manufacturing and professional interests are always represented in Parliament. Farmers never were. It's to their own interest to have a say in making laws which we all come under."[88] It was during elections that people were exposed to the tactics of the old parties, and to the rhetoric of central UFO officials. In the latter case, the words were absorbed, refined, and enhanced by local members.

First, however, it should be noted that UFO clubs had autonomy in deciding how candidates were chosen,[89] and the method sometimes changed on an *ad hoc* basis.[90] The fact that so many clubs tried to be equitable in delegate selection speaks well for the inclusiveness of the movement at that level. For instance, the East Lambton UFO ensured that each polling subdivision in the riding selected four delegates, two men and two women, for the 1919 election convention.[91] Local clubs also determined how the election campaigns were to be conducted and, more importantly, how they were to be financed.[92] In addition, local clubs determined whether or not candidates were obliged to sign recall papers.[93]

There is little doubt that the UFO leaders played a key role in the election campaigns. Farmers already knew or felt most of what they said, but having leaders from the central body speak at local meetings undoubtedly provided validation for many farmers by letting them know that they were not alone, and that their opinions were not unique. As early as 1917 UFO president R.H. Halbert declared: "Government by the people is a myth. The real rulers of Canada are the knighted heads of combines. Financial, manufacturing and food distributing interests are organized, and the individual farmer, standing alone, has no chance against them."[94] Halbert told a Simcoe audience in 1919 that the old parties would try to undermine the movement by attempting to nominate UFO members for the provincial election. He opined that any member who accepted such a nomination "deserved to get licked." A.A. Powers of the UFCC attacked leaders from both of the old parties as being lackeys of the Big Interests and claimed that "there is only one cure for this thing and that's to go into ... politics."[95]

Farmers were strongly urged to support UFO candidates because they came from their own ranks. West Simcoe candidate Richard Baker noted that, as a farmer, he knew the problems agrarians faced every day, and he instructed voters to vote for a lawyer if they wished to have an MPP who did not understand their concerns. A.E. Vance, a Lambton UFO member, echoed these views and in 1919 contrasted the number of farmers in the Legislature to farmers as a percentage of the provincial population. Indeed, there were elements in the candidates' speeches that struck responsive chords among the electorate. South Simcoe candidate Compton Jeffs informed an audience in 1919 that democracy had disappeared in Ontario during the war: "What democracy have we won at home when an officer whose commission may have come by pull receives double the pension of a private who fought in the ranks?"[96]

Of course, local members also had much to say, and they were often much more pointed in their comments than those nominated as candidates or those from the central UFO. Trevor Maguire of the Perth club asked farmers and labourers to unite, since both groups were "robbed at the point of production" by the Big Interests. G.A. Burgess of Carleton Place told farmers that he once supported the Tory incumbent in North Lanark, but now he was working to defeat him because "he was not progressive enough." The local "UFO Column" demonstrates the farmers' self-confidence and mockery of the old parties. In an article written shortly before the 1919 election, it was noted that several Liberal candidates were afraid to run in the contest, and that the Tories were so frightened at the prospect of losing that they were "throwing their old friends over board and scouring the townships for farmer candidates who are willing to lend themselves to the discredited party."[97]

As was the case in 1919, during the 1921 federal election campaign UFO leaders provided fuel for rank-and-file members with their rhetorical flourishes at local meetings. In 1920 at a Lambton UFO meeting, Drury's Agriculture minister, Manning Doherty, blamed industrialists for maintaining the artificially high prices: "The very fact that the big interests have been holding stubbornly back sustaining high prices, has put the buying public in a like attitude and I ... warn those leaders of industry and capital that no matter how strong and how numerous they may be they cannot successfully compete with the laws of nature." T.A. Crerar informed a Barrie audience that what was needed was unity in the country "and a vision of the big things. We should forget appeals to racial and religious prejudices and have the Canadian spirit, which existed during the war and which would make for purity, decency and economy in public administration." E.C. Drury claimed that the Progressive party was formed because "evils existed and ... the Government did not try to right them." Thus, the people had to do so: "That is the issue. The old parties are tied up, and our only hope lies in the new Progressive movement. It had started with the farmers, but it was not confined to them now."[98]

Rank-and-file members were indeed ready for a fight in 1921. Partly because the press was more sympathetic to the movement in

Lanark County than in either Lambton or Simcoe,[99] their feelings were recorded to a much greater extent there than in the other two counties. Hence the experience of Lanark UFO members is highlighted here.

The meetings held to nominate candidates in Lanark featured farmers at their most confident and most determined. A local newspaper reprinted comments made by various members at the 1921 convention, and some of them merit inclusion here. G.A. Burgess quipped: "I was a candidate in the 1917 election. R.L. Borden telephoned me and asked me to retire. I refused. If I had asked him to retire would he have thought it very cheeky?" William Code likened the political position of farmers to that of "Israelites in the wilderness, but we haven't had a Moses." Mrs. George W. Buchanan wryly observed: "I don't know about the House of Commons but there is the Senate. These old men need a woman to look after them." M.J. Smith declared: "No one can alleviate the condition of the farmer so well as the farmer himself. We congratulate ourselves that we are a free people, able to govern ourselves, but we have a government in power without a right to govern." For W.H. Robertson, "Canada's greatest drawback is that the wealth of the land has drifted into the hands of the few."[100]

In these words, one sees the contempt that members felt for those who had for so long had nothing but contempt for them. Now that farmers were organized, profound changes would be made; Canada would be the democracy it deserved to be.

Lanark Progressives selected R.M. Anderson, a "well-known and prominent farmer of Bathurst Township" who had been township clerk for some fifteen years before his nomination. Anderson took a militant stance during the election. In his advertisements, which differed from the other ridings only in the vehemence of the message, readers were informed that the tariff had

> fostered combines, trusts and "gentlemen's agreements" in almost every line of Canadian industrial enterprise by means of which the people of Canada – both urban and rural – have been shamefully exploited through the elimination of competition ... [The tariff] has been and is a chief corrupting influence in our national life because the protected interests have contributed lavishly to political and campaign funds thus encouraging both political parties to look to them for support, thereby lowering the standard of public morality.[101]

Anderson ran several advertisements during the campaign, but one, in particular, stands out. Under the headline "Manifesto of the

Smiths Falls Progressive Committee," the essay-length document outlined all of the injustices that "the masses" had endured. Asserting that unemployment and debt had led to a crisis in Canada, the advertisement asked who was responsible. "They are known as the practical hard-headed businessmen and educated politicians," the same people, it was argued, who failed to prosecute their friends who had been implicated in wartime scandals. In response to Tory and Grit claims that the tariff was the people's salvation, the Progressives warned: "Don't be fooled. Tariff cannot and will not cure unemployment. Why has it not prevented it? The causes are deeper and arise out of our present economic system and its methods of production. The exploitation of the people is the same whether you have high tariff or free trade." In fact, it was argued that the economic system was responsible for war, since tariffs led to retaliation by competing countries and such retaliations could escalate into armed conflict. It was even alleged that those in power wished to maintain this system, and "put forth every effort to blind you to the real issue."

On labour issues, the manifesto noted that the federal government had violated fair wage regulations, though it posed as a friend to the worker. Worse, there was misery in a land of plenty, yet politicians were "exposing for you the iniquities of their party machine with its vicious patronage operation, reckless expenditure of public funds, maladministration of our laws in respect to the payments of taxation, political appointments to the Senate by a decadent Government, and we are the victims."

Of course, none of this would be possible without the assistance of the Big Interests, who were placarding the country, "controlling the press, misrepresenting the true state of affairs ... [so] that they may again be elected to power for another period of exploitation." Reminding readers how they had been treated in the past by these people, the authors of the manifesto argued that neither the Conservatives nor the Liberals promised anything different this time around: "DON'T BE FOOLED ANY LONGER." From the present government, "all you can expect is words." Progressives, however, wanted action and called for a reduction in unemployment, elimination of profiteering, implementation of fair taxation, and better treatment for returned soldiers. It was clear that voters should support candidates who knew their problems, people from their own ranks "who are concerned in the welfare of the struggling masses of humanity, and not the favoured few."[102]

The manifesto is a striking document, not only for its length but also for its clarity and logic. The Big Interests at that point were not some abstraction; they were actual people who engaged in business and politics, and who perpetuated lies. Neither Grits nor Tories had alleviated misery; neither had produced full employment nor equality of opportunity. Moreover, real democracy was the furthest thing from their minds. The manifesto attests to the resentment that UFO members felt over the state of affairs in which they found themselves. Their alternative vision found concrete expression in a document produced not by head office but by a local group ratified by the mass membership. Moreover, it made sense to 6,615 of the 15,865 voters in Lanark who chose the Progressives in the 1921 election.[103]

These ideas resonated throughout the ridings. In Simcoe two UFO members appeared at a United Farm Women of Ontario meeting at Cooper's Falls, and one of them denounced John A. Macdonald and Wilfrid Laurier as traitors to Canadians.[104] In North Simcoe, UFO members discovered a speech delivered by Tory incumbent Col J.A. Currie in which he said that he had informed the House of Commons of his willingness to use machine guns during the farmers' march to Ottawa in 1917: "Just think of the courage that would be necessary to turn a machine gun on unarmed civilians! ... Yet regardless of this he tells us that the farmers will vote for him. He must think the farmers are easy. They will hurl his insults back in his face and vote for Ross."[105] In another article it was noted that the Big Interests realized that the defeat of the government of Arthur Meighen would mean the end of "industrial piracy." Thus, the "Big Interests will spend millions of dollars to prevent the defeat of this paternal government."[106] Undoubtedly, there was more at stake in the election than the tariff, prohibition, or rural representation in the Legislature.

Progressive candidates repeatedly validated the thoughts of rank-and-file members. East Simcoe candidate Thomas Swindle appealed to those who had supported the old parties in the past. He realized that it was difficult to break away from old party loyalties, but if people did so they would be rewarded: "When you vote for a Progressive candidate you vote for a government by the people and for the people." Swindle also claimed that the Progressive party emerged out of the failure of the old parties "to play square with the common people." As a result, "independent political action seemed to be the only course open to them." In fact, Swindle advanced a platform similar to that of the Lanark Progressives.[107] North Simcoe candidate Thomas E. Ross believed that the central issue of the election was whether or not Canada was to have "government by the people for the people or continue government by the interests for the interests." Ross was pleased that the Progressives did not have a central campaign fund: "Anyone who stoops to use such funds sells himself into bondage. Only by having representatives who will stand true to their principles, no matter what pressure is brought to bear upon them, can Canada be extricated from its present unfavourable condition."[108]

DEVELOPING ALLIANCES IN IMPLEMENTING THE UFO VISION

Given the enthusiasm and evangelical bent of many UFO members, it is understandable that they sought alliances with other groups with similar objectives. The most logical ally, it seemed to many members, was the labour movement, which too was in the process of experimenting with political action with the newly formed Independent Labor Party (ILP).

In most accounts of the UFO, the explanation given for the division between labour and farmers is that labour was pro-tariff while farmers were decidedly not.[109] This interpretation is borne out to a certain extent when the central organizations of the two groups are examined, but when the focus moves to the grassroots, it proves too simplistic. The tariff, although a concern, had little to do with local battles between farmers and labourers, as UFO/ILP relations in Lanark, Lambton, and Simcoe counties demonstrate.

It appears that local UFO and labour groups had little to do with each other before the 1919 Ontario election. When the election was called, it seems that local UFO and labour organizations came together spontaneously to fight what they perceived to be a common foe – the Big Interests. It was once they had come together that some profound differences emerged between the two groups.

In several ridings the UFO invited labourers and returned soldiers to nomination meetings, and both eagerly accepted.[110] In some ridings, such as West Simcoe, relations between farmers and labourers remained fairly cordial for the duration of the campaign. The only divisive issue was that of the eight-hour day. Richard Baker, the UFO candidate for the riding and a farmer, avoided a split between the two groups by stating his support for the eight-hour day, as long as it did not apply to farm labour. The main union in the riding was the Collingwood Maritime Trades Association, which backed Baker's campaign.[111] In others, UFO overtures for a united campaign were rebuffed. In North Lanark, for example, UFO members approached the Almonte local of the United Textile Workers of America in order to obtain its support. Farmers pointed out that they received nine cents per pound for prime beef, while workers paid forty cents per pound for steak; the two groups should cooperate to reduce this gap.

In response, the union claimed that it had too much on its agenda to consider the proposal.[112]

In several jurisdictions no effort at all was made to unite farmer and labour forces officially. This was the case in East Lambton, where the prospect of meeting with ILP officials does not seem to have been raised during meetings held to determine whether or not to field a UFO candidate.[113] Elsewhere, confusion resulted in UFO and ILP candidates squaring off against each other in the election. This occurred in South Lanark, where the local UFO nominated a farmer candidate without consulting with the ILP. In response, the ILP nominated Richard Grant as its candidate. UFO candidate W.I. Johnson suggested that, if labour supported the UFO in the provincial contest, the UFO would return the favour by backing an ILP candidate in the next federal election, but to no avail.[114]

East Simcoe provides a good example of the difficulties that had to be overcome in order to field a UFO/ILP candidate. During the nomination meeting, to which both returned soldiers and workers were invited, Frank Foster, business manager of the Orillia Federal Labour Union (which, according to Foster, had 652 members), stated that the only plank that soldiers and labourers could not endorse was the call for prohibition. The issue, Foster believed, should be left for the people to decide in a referendum. Not even the farmers' stand on the tariff would cause much trouble, said Foster, who thought that high tariffs did not help Ontario workers in any meaningful way.

When asked if there were any aspects of the ILP's platform that the farmers might object to, meeting chairman E.C. Drury cited the eight-hour day. Drury asked labour and soldier delegates to put themselves in the place of farmers, who could never work an eight-hour day and expect to run a profitable operation. Moreover, if urban workers were granted an eight-hour day, it would make towns and cities more attractive to rural youth, providing more incentive for them to leave the farm when the labour situation was already dire. Instead of offering to support the plank, Drury argued that the only difference between farmers and labourers was that workers sold their labour for wages, whereas farmers sold theirs "in the form of produce." Subsequently the delegates were presented with three options: farmers could nominate their own candidate; farmers could support an ILP candidate; or a joint UFO/ILP convention could be held to nominate a joint candidate. Ultimately the third option won out, and a committee was struck to devise a mutually satisfactory platform. Those in attendance agreed to meet again the following month.[115]

Approximately 450 delegates attended the September meeting in Orillia: two hundred UFO members, two hundred ILP supporters, and fifty returned soldiers. Drury reported that the ILP and UFO platforms had been effectively combined with a few alterations, which went unreported. Before proceeding with the selection of the candidate, he made two things clear. First, whoever was chosen would have the full support of both the UFO and the ILP, no matter who won, and, second, the candidate would be required to sign recall papers. John Benjamin Johnston, a manufacturer's agent for stoves and milking machines, was nominated as the UFO/ILP/ soldier candidate.[116] Despite substantial opposition from many local newspapers, the fragility of the alliance, and the candidate's limited talent, Johnston defeated his Conservative and Grit opponents.[117] In East Simcoe, farmers, labour, and soldiers had successfully united to defeat the Big Interests.[118]

Despite the victory and the coalition government, difficulties in reconciling the three groups remained. The 1920 annual UFO convention in East Simcoe featured a prolonged debate on whether or not to form a united front of farmers, workers, and returned soldiers to contest the next federal election. Trades and Labour Council representative A. Jackson submitted four planks that the council wished to see added to the Farmers' Platform: an eight-hour day; old age pensions; a fair rent clause; and inclusion of profiteering in the Criminal Code. Thomas Swindle, chair of the meeting, felt that the only contentious point was the eight-hour day, but in a conciliatory gesture he said that the Progressives could suggest to Parliament that a board be appointed "to fix the hours in each industry ... making the health of the men the first consideration." This appeased the labourites, who agreed to work out a mutually satisfactory platform.[119]

Tensions between the two groups in East Simcoe mounted, however, as the election neared. UFO, ILP, and returned soldier delegates – more than seven hundred in all – gathered at Midland in October 1921 to select a candidate. There had been rumours that the ILP would work for the nomination of Manley Chew, a local manufacturer and former federal Liberal candidate, as the Progressive candidate.[120] Why labour supported Chew is unknown (some ILP members may have believed that he would attract Liberals to the Progressive cause), but several farmer delegates adamantly opposed his nomination, insisting that an arrangement had been made in 1919 whereby the riding was to be represented provincially by the ILP and federally by the UFO. The agreement seems to have been lost sometime between 1919 and 1921 but allegedly there was copy that read: "We, the undersigned, on behalf of the United Farmers, returned soldiers and Laborites, agree that the following shall be the basis of representation: Labor and Returned Soldiers to have representation in the Ontario Legislature, and the UFO to name a candidate for the Federal House, to be endorsed by the aforesaid body and supported to the last ditch." The document, dated 27 September 1919, was supposedly signed by twelve prominent Labourites, UFO members, and returned soldiers.[121] Tempers flared to the point where one of the UFO organizers told the farmer delegates to adjourn to a nearby hall to nominate their own candidate, since the UFO could not accept a manufacturer as its representative.[122]

As the farmers rose to leave, local labour leader David Kennedy announced that there was still hope that a compromise could be reached, "as the Labor Men were in a position to nominate a candidate who has the confidence of every labour man, farmer and soldier who had ever done a cent's worth of business with him," meaning Chew. Kennedy also said he was unaware that anything in the UFO constitution prevented a manufacturer from running as a UFO candidate. Thus, he reasoned, the objection to Chew "was not one of principle." He then led the Labor and soldier delegates out of the hall so that the UFO delegates could determine how to proceed.

Now on their own, the farmer delegates discussed what had just transpired. G.R. Murdoch, UFO MPP for Centre Simcoe, expressed his regret at what had happened but said that the incident was not the farmers' fault: "The whole thing had been organized from beginning to end by capitalistic interests ... The farmers were for equal rights for all, and privileges to none. The capitalists were for themselves." The ILP delegates, Murdoch contended, were just as self-destructive as those in the Legislature. J.B. Johnston, the newly elected UFO MPP for East Simcoe, did not think that the rift between the ILP and the UFO would prove to be permanent. He referred to himself as a representative of the workers and thought that the delegates should simply select the candidate with the greatest number of votes. At that point, a Mr Yeats of Penetanguishene said that when Johnston was nominated he had signed an agreement stating that a farmer should have the federal nomination, to which Johnston said that he had "never repudiated the agreement." Thomas Swindle then stated that the accord had been drafted between the ILP and the UFO, and that it "gave the farmers the right to have a farmer candidate, and Labour had no say." He went on to recall that, at the time the accord was signed, the workers had said, "Put up your farmer and we will show you how we stick." Of the present dilemma Swindle said he was glad that "the farmers had stood up and said no. They wanted no manufacturer or manufacturer's tool" to represent the Progressives. Frank Foster argued that there was enough talent to choose from among the farmers, so there was no need to look elsewhere. Although Chew might appear to be a worthy citizen, farmers "had been beguiled before." Chew, Foster claimed, was "one of the capitalistic class. If elected would he be true to [the] principles" of the UFO?

At the end of the day the farmers decided to nominate their own candidate. Thomas Swindle was chosen, and he sent a number of olive branches to the ILP and returned soldiers. He stated, for example, that in principle he favoured an eight-hour day (although it should not be legislated until later), and that he felt that the government should pay returned soldiers a sum equal to what they would have received if they had been employed at home during the war.[123] The ILP never officially responded to these overtures.

Living in a system in which the interests of workers were at odds with those of farmers, it is natural to assume that there would be differences between the two groups in their fights against the Big Interests. In many cases, the differences were too profound to overcome. But in some ridings there were sincere and creative efforts to unite agrarian and worker parties in an attempt to gain representation in the Legislature.

CONCLUSION

The Great War and the period that followed contained many of the elements necessary for the emergence of a movement culture of opposition. During the war, farmers saw government try to take away their labour force for military service, while simultaneously granting benefits to industrialists who profited from the conflict. For many UFO members this was yet another example of a polity dominated by the wishes of the Big Interests. The press, itself part of the dominant class, continued to mystify Canadians about the true state of affairs with its use of propaganda and deception. Urban society, growing at an alarming rate, was pushing its own version of morality upon the populace, a morality that was at odds with what most farmers believed.

Where, then, did UFO members see themselves in terms of class and in relation to capitalism and the free enterprise market system? Clearly, they saw themselves as part of an exploited group. Without a monopoly over agricultural produce (which was not, I argue later, what they wanted), farmers were vulnerable to the decisions of food-processing companies and retailers, who were in turn conditioned by international markets and prices. In short, farmers were so far down the chain of production that their agency was negligible.[124] UFO adherents also believed that they had no say in the political process. One discerns in the movement an attempt to eliminate the power imbalances that it perceived.

Hope, idealism, and enthusiasm characterized the movement in 1921 because of increasing citizen participation in public policy formulation. One sees in the UFO a strong, if unconscious, strain of anarchism. It became clear during the immediate postwar period that something was wrong. The traditional belief that hard work and moral living led to a rich and contented life was exposed as a myth. These values in themselves could not produce a harmonious society given the larger forces that sought to destroy it. It was only through the diligent efforts of groups such as the UFO that the Big Interests had not yet succeeded in doing so. Still, it was demoralizing to have to be so vigilant of the society one lived in.

Local UFO clubs banded together to fight those whom they held to be responsible on their own terms. Rather than be slaves to the dictates of middlemen, they established cooperatives. In response to what they saw as a hopelessly corrupt system, they decided to contest elections. That they were able to produce a political force capable of securing power provincially and electing enough MPs to be the official opposition (though the right was not exercised) was a formidable accomplishment. UFO members had done their part; it was now up to those they elected to begin the process of setting things right.

4 "Sane and Careful Administration": The UFO Movement in Decline

Rank-and-file UFO members might be excused for believing that Canadians would witness profound change after the 1921 federal election. After all, the UFO had formed the government in Ontario in 1919 and the Progressive party, having just contested its first election, now occupied sixty-four House of Commons seats (twenty-four of which were from Ontario ridings) – as noted earlier, enough to form the official opposition. Equally important, in the 1919 provincial election the UFO received 21.7 percent of the popular vote. In 1921 the Progressives received 27.7 percent of the popular vote in the province: there was evidence of momentum at that point.[1] Many farmers took the words of the movement's leaders at face value and envisioned a new society that would check the influence of the Big Interests and give common people more agency in the governing of their country.

Aside from a few reforms, however, the promise of change did not materialize in any meaningful sense, and by 1923 the UFO had entered into a period of decline from which it never recovered. This chapter explores how and why this happened. Although some members, and even some clubs, tried to advance an alternative vision of what society should be like, their efforts were negated by the central body and UFO legislators who, for several reasons, moderated their stances once elected. At the federal level, reforms were virtually impossible to effect because the Progressives were reluctant to use their power to defeat the Liberal government of William Lyon Mackenzie King. At the provincial level, UFO MPPs were unwilling to rock the boat in any significant way, and the "broadening out" controversy – the split over whether the UFO should expand so as to include non-farmers or whether it should remain an agrarian movement – sapped the UFO of much of its energy. One may debate the causes of the UFO's decline – inexperience, strained relations with labour, deference to power, internal discord, and so on – but its consequences are indisputable. The widespread changes that rank-and-file UFO members expected never materialized.

There was more to the UFO's defeat than its declining ties with labour and its parliamentary failures. The UFO did not exist in a vacuum; it was part of a society in which a certain outlook was so firmly entrenched that it was accepted almost as revealed truth. That the movement was able to challenge some of its society's fundamental assumptions stands as testament to its perseverance, and to its ability to critically evaluate what surrounded it. Yet the pressures wielded by the "old order" made it extremely difficult to sustain this effort.

In Ontario there were those who benefited – sometimes immensely – from the preservation of the *status quo*. Moreover, although the UFO attained a measure of political power, those who opposed it still had other forms of power at their disposal that they could deploy. The old order was not going to pass away without a fight, and the strategies it employed in combating the UFO were numerous. The movement could be ridiculed, discredited, ignored as a statistical aberration,[2] or, even more effectively, dismissed as well-intentioned but wrong-headed. A portion of this chapter is thus devoted to the power of hegemony, and to those cultural agents that either actively promoted the benefits of returning to the old order or recaptured UFO members support through a host of strategies.

UFO members faced a constant barrage of mixed messages from their leaders, the media, the old political parties, and, at times, from within their own clubs. Ideally, each of these agents should be treated as a discrete entity, but it is more realistic to describe how they intersected and reinforced one another to bring the full weight of hegemony to bear on the UFO rank and file.

Describing the decline of a movement such as the UFO requires some speculation. Yet if one reads enough of the evidence, one discerns that local members became demoralized as they gradually realized that their monumental effort to break free of repressive societal norms did not automatically bring about fundamental change. It is in this context, then, that one obtains a better understanding of how UFO members became resigned to the *status quo* and of why many of them returned to old political allegiances.

UFO LEADERS

The standard interpretation of the UFO is that it was, in the main, "crypto-Liberal" in thought and practice. As the last chapter showed, this generalization is misconceived since it paints all UFO members with the same brush. The misconception is understandable, however, given the attention that has been paid to the movement's leadership, where the UFO's crypto-Liberal stream found its clearest expression. Since the leaders were often portrayed and perceived as the force that would liberate farmers (and, by extension, society) from serfdom to the old parties and vested interests, the implications of their crypto-liberalism are profound.[3] Examples of the leaders' thoughts and actions after obtaining provincial power and federal prominence serve not only to illustrate the importance that these figures were accorded by the media and by UFO members, but also to demonstrate the influence that the messages they disseminated ultimately had on the movement's rank and file.

Owing to his position as premier, E.C. Drury was closely watched by the press and by the UFO's rank and file. Drury realized that he could not afford to alienate members of his support base, and he genuinely tried to address some of their concerns. But there were indications early on that farmers would have to be vigilant of Drury and his ministers.[4]

Immediately after the 1919 election victory, Drury, although premier, did not have a seat in the Legislature, nor did Agriculture minister Manning Doherty or Attorney General W.E. Raney. Tempers flared in the riding of South Renfrew, near Lanark, when it was rumoured that UFO MPP John Carty might be asked to resign his seat in favour of Doherty. The central UFO eventually chose another riding, but the incident left some local members wondering how different the new government was from its predecessor: they had almost been asked to sacrifice constituency autonomy – and their MPP – to central authority.[5] Aside from a few slips like this,[6] however, Drury convinced many supporters that he was leading "a movement of Democracy," to use his own words, that would enable common people to participate meaningfully in the political process.[7]

Over the next three years, the Drury government legislated, by contemporary standards, several progressive measures, but little changed regarding the political process.[8] Although Drury boasted during the 1923 campaign of having implemented several reforms, he was forced to remain silent on various elements of the farmers' platform.[9] An examination of how Drury conducted the campaign provides a good example of the mixed messages UFO members received.

In May 1923, shortly before the election, Drury drew a packed house at the Balderson Theatre in Perth. He began his speech by noting that it was "a wholesome thing when the citizens could turn out at even a busy time to hear public matters discussed." What followed, however, was not so much a discussion as a lecture. He spoke at length in defence of his government's accomplishments, particularly those concerned with temperance legislation. Drury argued that the UFO should be re-elected because his government "stood for sane and careful administration, bettering of the educational system, safe Hydro Electric and Hydro Radial development, bettering administration of forests and enforcement of the Ontario Temperance Act." All noble goals, but they were not in the same league as promising to quash the Big Interests and eliminate the stranglehold they had on Ontario's citizens, the commitment that had been made in the previous campaign. In fact, if the press accounts of his speech are accurate, Drury made no mention of the importance of checking the influence of the Big Interests, the desirability of reforming the electoral system, or the need for greater citizen participation in public affairs.[10]

During the 1923 campaign Drury was forced to adopt a defensive position; after all, he had to campaign on the performance of his government. Although he easily defended his record on radials, temperance, forest management, and similar moderate accomplishments, other policies created problems for him. Instead of outlawing racetrack betting, for example, the government had imposed a tax on winnings. Justifying the policy in Simcoe during the campaign, Drury argued that, while drinking could be abolished, "the Province could not stop betting, so what was wrong in taking a part of the proceeds and making it a little less attractive."[11] Given the strong evangelical tendency within the UFO and its hope for inculcating the polity with a greater sense of morality, this position – from a government that had enacted some of the most stringent temperance legislation in the country – must have perplexed and even annoyed some supporters.

During the 1923 campaign Drury was largely silent about such matters as the transferable vote and proportional representation.[12] Conversely, he stated that if he and his government were returned, he would implement some form of closure in the Legislature. He had opposed such measures in the past, but he had changed his views and now felt that the government needed a means of ensuring that debate ended after a reasonable time. According to Drury, the filibuster that had been staged at Queen's Park shortly before the election was called meant that the 1923 contest was being held "without redistribution and without provision for the transferable vote."[13] Thus, instead of advancing a policy that allowed for expanded democratic forms, Drury used part of the campaign to promote a contrary measure – reducing the freedom of MPPs in the Legislature. As a representative of a party that advocated constituency autonomy and MPP responsibility towards their constituents, this must have been confusing for members.

In August 1923, after the movement's electoral defeat, Drury announced that he would not continue to lead the UFO unless a convention was held to ratify his leadership, and unless he was free to pursue his plans for a People's party. In a speech he delivered at a UFO picnic at Wasaga Beach later that month, he actually attacked the UFO, claiming that it was run by a clique. What was needed, he declared, was for the movement to broaden its base of support. Although it is not known how this would eliminate élite domination of the UFO, Drury's sentiments "appeared to find favour with many of the hearers, to judge by the applause."[14]

That Drury drew applause for his proposed People's party underscores a significant problem facing rank-and-file members: what was the UFO's ultimate goal? For some, the movement existed for the benefit of farmers alone. If other groups wished to organize to safeguard their interests, and if those interests sometimes intersected with the interests of farmers, then well and good. Such groups, however, were to remain separate and distinct from the UFO. There were other members, conversely, who saw the UFO as a movement that should ultimately benefit all common people. If the democratic measures it advocated were implemented, then other groups would be able to liberate themselves as the UFO had, which would lead inevitably to the betterment of humankind. Thus, there was nothing wrong with forming a People's party – that was the logical conclusion of the UFO's agenda. Still others appeared to make no distinction between the two camps; they seemed either to support Drury and J.J. Morrison at the same time or to dismiss the debate as irrelevant to their locally based democratic agenda. As will be seen, the issue was never satisfactorily resolved among the movement's rank and file.

Drury continued to have a substantial effect on the movement, or at least upon Simcoe UFO members. After losing his seat in 1923, he sought federal office in the 1925 contest. He was the only Progressive candidate for the two federal Simcoe constituencies. Drury sought the North Simcoe seat, whose incumbent, T.E. Ross, was also a Progressive. Ross, it appears, intended to seek re-election, and at the nomination meeting he and Drury were both nominated to stand as candidates. In the end, however, Ross refused to run, claiming that Drury's "proper sphere was in Federal politics."[15] It is not known whether Ross reached this decision by himself or whether, intimidated by Drury's status, he felt compelled to step aside. In any event, his withdrawal undoubtedly disappointed many of his supporters, even if their new candidate was a former premier.

Drury ran an uncontroversial campaign. In his view the key issue was the tariff, and he devoted himself to advocating its elimination. The rhetoric he used is revealing. One of his speeches in particular provides a rare glimpse of his views on what Progressivism stood for: "The question of protection is a great moral one. Allowing one class to take from another for their own personal benefit is a great moral question ... We are the only party interested in the welfare of the common people."[16] While Drury may have seen the tariff as the primary source of the country's problems, more often than not he failed to make the connection clear to his audiences. Regardless of his sincerity, the fact remains that he did not delve to any great extent into the consequences of protection. Most notably, he made scant reference to the Big Interests and how they benefited from such measures as high tariffs at the expense of the common people.[17]

Drury was not the only UFO leader to propagate moderate or mixed messages after the electoral successes. R.H. Halbert, speaking to Almonte UFO members, said that before he was elected to the House of Commons he believed that the person who stole from the public purse was the greatest danger to the country. Since being elected, however, he had changed his mind and now believed that "the man who stirs the fires of racial and religious hatred is the greatest menace the country had today." That Halbert, who at that time was an Orange Lodge member,[18] could become more tolerant is admirable; that he now believed racist forces to be more dangerous than the Big Interests is more problematic. Some UFO members believed that the Big Interests actually fostered racial hatred and various forms of intolerance to suit their own ends. Thus, Halbert may have presented his audience with a cause-and-effect inversion that made little sense.

Even J.J. Morrison, the leader most often labelled a radical, did not always act accordingly. At a meeting in Orillia in late 1922, he identified a problem facing Progressive MPs. According to Morrison, they were reluctant to defeat the King government for fear of forcing an election. To do so meant risking electoral defeat: "If you were in that position you would sidestep too. The personal interests of the Member of Parliament come into conflict with their duty and it is difficult under those circumstances to get good government."[19]

Morrison may have been alluding to the futility of parliamentary democracy, but another message can be discerned: Progressive MPs could be forgiven for sacrificing good government to self-interest; it was human nature to do so. Morrison's comments could be taken as an insult to UFO members who, he was implying, would also sacrifice principle for power if they were faced with a similar situation.

Speaking in Lanark County during the 1923 provincial campaign, Progressive leader Robert Forke spoke in glowing terms of the accomplishments of the UFO/ILP coalition. He also expressed his hope that Drury would retain the premiership, if for no other reason than that a defeat would have a negative effect on the national movement. Even if Drury were to lose, Forke argued, the farmers had already altered the political landscape in Canada forever: the election of the Liberals or Conservatives would not make a great deal of difference because "the influence of the Progressive movement will live." In effect, Forke gave the farmers license to withdraw their political support from the UFO. He extended this thought one step further by reminding his audience that the movement was "more than a political [one]. It is a reaching out towards an ideal."[20] Not long after, however, Forke seemed to contradict himself. Apparently forsaking ideals, he told an audience in a nearby riding that his experience in Ottawa had taught him the value of compromise. He urged his audience not to put too many planks in the party's platform because, "although certain principles never changed, yet in political life circumstances and conditions often altered."[21]

The foregoing is a brief survey of the sorts of messages transmitted to the UFO rank and file once political power was attained. Undoubtedly, the leaders were faced with circumstances that necessitated conventional approaches. The UFO government had to address issues that went beyond purely local matters; radials, trade, education, resource management, and so on required action. Moreover, they needed to be dealt with through an established bureaucracy and established practices. Indeed, elected politicians were only one component of a complex political structure that had been legitimized and firmly entrenched in Ontario and throughout Canada.[22]

The constraints were profound, but there was room to implement measures to loosen the boundaries. But instead of attempting to implement the required changes, the leaders shifted their rhetoric. Largely absent after 1921 was the oppositional stance that they had originally taken against the Big Interests and the inherently corrupt and undemocratic political system that bolstered them. Now the best the leaders seemed capable of doing was to promise "sane and careful" government with modest reforms. Although this pragmatic position may have represented all that they thought they could realistically achieve under the circumstances, it almost certainly had a negative effect on the idealistic UFO rank and file.

Yet there was room for optimism. Between the UFO élite and the mass membership were the UFO/Progressive MPPs and MPs who, because they were drawn from the "common people," were given the task of going off to their respective legislatures and fighting to have the will of their constituents heard and implemented. If they remained true to the principles of their supporters, then there remained reason for hope and, indeed, idealism, among the mass membership.

There were relatively few instances when UFO/Progressive MPPs and MPs incurred the wrath of local members. There were times, however, when members must have wondered exactly what their representatives were doing in Toronto and Ottawa. At a Ramsay Farmers' Club picnic in 1921, North Lanark MPP Hiram McCreary discussed the bonus that MPPs had voted themselves that year. After declining to accept the $2,500 sessional indemnity in 1920, Drury suggested that a $600 bonus be paid to MPPs for 1920–21. McCreary told his audience of his response when the proposal was put to him: "I said nothing at all against it. I was willing to take it, and I was willing to do without it, but I don't believe in a man holding up his hand against a thing yet knowing that he is going to get it, and then coming back to his constituents saying 'I did not vote for it.' If there is any blame to the government for taking it, I stand or fall by it. I voted for it." Although such honesty and courage, one paper opined, would win McCreary new supporters, some UFO members must have wondered why the bonus was considered in the first place, given the relatively high rate of pay MPPs received. East Simcoe UFO MPP J.B. Johnston was in favour of paying MPPs higher wages. He reasoned that higher pay might induce good, free-thinking people to run for provincial office. Local UFO members, such as Donald MacPhail of Perth, did not agree. In fact, MacPhail wished to see MPP's salaries reduced. In his view, if farmers had to economize, then so should elected officials: "If the present Government does not take the opportunity [to reduce salaries], I am sure no other Government will."[23]

Despite the rhetoric during election campaigns, most UFO/Progressive MPPs and MPs behaved conventionally once elected. This can be explained, in part, by the fact that the existing political system demanded such behaviour, but there were other reasons as well.

UFO MPPs found that their tasks as representatives were more complicated than they perhaps had first envisioned, and that it was easier to revert to old party tactics than it was to sustain a radical position or even to provide straightforward explanations for their actions. Hiram McCreary, for one, became proficient in the former approach. Upon being nominated to contest his seat in 1923 during a UFO/ILP convention, McCreary boasted at length of the Drury government's accomplishments, including the removal of Great Britain's embargo on Canadian beef, the encouragement of cooperative marketing of agricultural produce, and the construction of provincial highways. He took personal credit for having a highway pass through Carleton Place rather than through Smiths Falls, which was outside his riding. Instead of receiving applause, however, McCreary was asked by one audience member: "Why did you not get it through Almonte?" In traditional roads-for-votes style, McCreary replied "I'm going to get something for you yet. We're working on it now. Send the right people to Parliament and you'll get good roads."[24] Avoiding any promises to implement an alternative electoral system or to improve democratic mechanisms to check the power of the Big Interests, McCreary adopted, in his response, all the trappings of an old-style political speech.

McCreary was not the only UFO politician to engage in this sort of behaviour. East Simcoe MPP J.B. Johnston boasted that the riding received more government grants during his first term than it had "during the whole Conservative regime." Johnston also promised that the Orillia-Penetang highway would materialize if he and the UFO were returned to power. North Simcoe Progressive MP T.E. Ross somewhat uncharacteristically informed an audience during the 1925 federal election that he had recently managed to obtain an allocation of $20,000 to repair and dredge the Collingwood Harbour, and that during his tenure as MP he had managed to secure a total of $168,000 for that purpose.[25]

To return to McCreary, although he was touted by the local press and his own advertisements during the 1919 election as a candidate of the people who would combat the Big Interests, he performed unspectacularly at Queen's Park. McCreary's time in office was noteworthy only for his introduction in 1920 of a motion requesting that the federal government hold a referendum on the importation of liquor into Ontario. For the most part, he performed the duties of a backbench MPP of the governing party. In 1923, with the provincial election call, McCreary again invoked the image of the UFO as the sole democratic party in the Legislature.[26]

McCreary's appeal to voters provides an example of how the rhetoric of earlier times persisted, despite conventional behaviour at Queen's Park. In one of his advertisements, readers were warned that "Democracy is on Trial"; hence they needed to support McCreary, "the Candidate of the People." In the same advertisement it was argued that the Drury government's social legislation eased "the strain on the body and mind of the worker," and that people should vote UFO so as to "Keep Democracy in Power."[27] The only problem with making these claims in 1923 was that, despite four years in office, McCreary could not show his constituents any significant change in how things were done in government. He was not alone in this; virtually every UFO MPP used the same sort of tactics during the 1923 election and displayed the same discrepancy between rhetoric and action.

Although populists want their elected officials to be "of the people," they also respect those who mature while in office and who are able to take on opposing politicians as equals. Of necessity, most UFO/Progressive MPPs and MPs became conversant not only in parliamentary practice but also in contemporary political tactics. Not to do so meant endless humiliation at the hands of members of the older parties. Yet to acquire source of the characteristics of one's opponents, however necessary, carried with it significant problems. One might begin to sound like those whom one professed to oppose.

By 1925 the rhetoric of opposition had shifted markedly. The UFO now spoke of providing "common sense" government rather than of serving as the bastion of democracy. The 1925 federal election campaign in Lanark illustrates the point.

UFO members in Lanark decided not to field a candidate in the 1925 election, probably because Duncan H. Gemmell, long known for his UFO sympathies, planned to run as an independent. A prohibitionist (an executive member of the local Prohibition Union), Gemmell said that the issue would play a significant role in his campaign.[28] As predicted, Gemmell received the support of the Lanark County UFO Political Association. In a pro-Gemmell advertisement John H. Chapman, the association's secretary, noted: "Whereas the political situation in Lanark County has materially changed since our last meeting, Mr Gemmell ... having entered the field as an independent candidate, and having announced his platform, which meets with our approval, we take this earliest opportunity of endorsing him as one who would be a worthy representative in the House of Commons, and would request the entire group of United Farmers to give him their hearty support."[29]

Gemmell claimed that he had decided to run as an independent because he was "firmly of the belief that the people ... of this constituency have reached that stage in their political life when they do not intend to be further exploited just for party's sake, but rather that their public wants and conditions would find expression in keeping with their spirit of democracy ... these things force me to the conclusion that there is still room for men of independent minds to render public service."[30] As for Gemmell's position on the issues, he believed that tariff matters should be left to legislators, not to a board, as some farmers proposed. He supported increased "yet selective" immigration and reform, not abolition, of the Senate. In announcing his platform, Gemmell asked for support on the grounds that he would work hard for all classes of citizens, and that he knew of the conditions of the common people.[31] Gone was any reference to the Big Interests, or any promise to expand democratic mechanisms. Instead, Gemmell merely pledged to work hard for everyone in the riding. His mild-mannered campaign – particularly his apparent refusal to address power inequities – may account in part for the weak support he received from the electorate.

Why Lanark UFO members, who displayed fierce opposition to conventional political practice in 1919 and 1921, supported Gemmell on this and other occasions instead of putting forward more militant candidates is not known.[32] It may have been because of the defeat suffered by R.M. Anderson in 1921, and the less-than-stellar performances of W.I. Johnson and McCreary, which might have further demoralized the rank and file. When combined with the factors discussed below, demoralization as cause becomes a distinct possibility.

Some UFO candidates who consistently questioned existing political practice and presented an alternative vision enjoyed success even after many of their more moderate colleagues suffered defeat. East Lambton MPP L.W. Oke maintained his position as a foe of the Big Interests and was returned in 1923 and 1926.[33] During the 1925 federal election, he alleged that the Big Interests encouraged rural depopulation and were consequently leading the country to ruin. Oke stated that the province could not

> have Toronto doubling its population every 15 years ... and us helping to build industry there, and every township, practically in the whole province losing in population – 5,000 in this riding alone. What is going to become of us nationally? And yet, when the farmer tries to get his own rights ... men of every calling in the industrial centres come out and decry the farmers' movement and his voice in public affairs. This is only stimulating us to fight for our rights, and what we believe to be in the best interests of the country.[34]

Whatever the merits of Oke's analysis, his assertion that farmers were fighting powerful forces from urban centres that were indifferent to their fate certainly struck some members as true. Doubtless, other factors contributed to Oke's success: he represented a predominantly rural riding, and he was supported by the local Prohibition Union in the 1926 provincial contest.[35] But in similar ridings in Ontario many voters had already returned to the old parties. It is reasonable to assume, then, that Oke's campaigns – both the views he expressed and his commitment to action – struck a responsive chord in many East Lambton farmers.[36]

THE CO-OPTING OF LOCAL CLUBS AND RANK-AND-FILE MEMBERS

As Lawrence Goodwyn noted, it is "overarchingly difficult" to build and sustain a mass democratic movement of opposition.[37] This certainly was the case for UFO members who had to contend not only with cleavages among the movement's leadership but also with the negative public perception of what the UFO hoped to accomplish. Yet there were those in the movement who persisted in advancing an agenda of change. Some clubs had sufficient numbers of these members to allow them to pursue this course consistently. In other instances, individual members continued to fight the moneyed interests, and all the political trappings that accompanied them, on their own.

Despite attacks from a largely hostile press, the machinations of the old parties, squabbles among the UFO leadership, and a host of other factors, there were rank-and-file UFO members who rejected the pervasive message of possessive individualism and continued to adhere to the original spirit of the movement, people who felt that there was more at stake during elections than mere promises to provide good government.[38] Their persistence stands as a tribute to a movement that could, at least for a time, politicize people with startling success.[39]

The independent attitude of local clubs manifested itself early on during the UFO government. Not long after the 1919 election, members reacted angrily to the central organization's proposal to increase membership fees. The Lambton County UFO, which represented seventy-five clubs, voted against the proposed increase, arguing that it was "unnecessary and harmful to the organization." Heber Shaw, a UFO member who lived near Lanark, summed up the rank-and-file side of the debate by writing that an increase in membership fees was the last thing the UFO needed. With the organization in power, he argued, the task was to orchestrate a mass movement by increasing membership. Higher fees would accomplish the exact opposite. Shaw also pointed out that, in farm households where there were two or more potential UFO members, some farmers might think twice about enrolling their sons, wives, or daughters.[40]

Several clubs tried to remain politically active. The Oro Station UFO, for instance, corresponded with its MPP on a regular basis.[41] The West Lambton UFO executive kept in contact with the UFO central office, corresponded with its MPP, and sent advice regarding legislation. In fact, the executive passed a motion stipulating that "when the member [could not] accept the advice of the riding executive he should give some reason why."[42] The Ardtrea UFO club in Simcoe invited the entire Orillia Township council and a provincial official to one of its meetings to discuss proposed changes to a nearby provincial highway. Later, the club invited a Department of Education official to discuss consolidated schools, and another meeting featured the chairman of the Orillia Water, Light and Power Commission. Federal and provincial politicians were often invited to speak to local UFO clubs on political matters.[43]

In East Lambton there was a concerted effort to follow the actions of the riding's MPP and MP, and to keep abreast of politics in general. Even in non-election years the political organization met in order to keep interest high. As elsewhere, politicians were often invited to speak to the group. Although some of their speeches were little more than reviews of the past legislative session, they did at times address specific issues, such as the broadening out issue.[44]

An examination of the East Lambton UFO Association shows how rank-and-file members persisted in their adherence to UFO principles. In April 1923, on the eve of the provincial election, the association met to plan for a May nomination meeting. As had been the case in the past, each polling subdivision sent two men and two women to the convention. Urban and rural UFO supporters had equal status and an urban resident could be nominated, "provided, however, that the candidate be in accord with principles" of the UFO. Not wanting their candidate to become a lackey of any party, members of the association also agreed that if the candidate was elected and the UFO formed the government, "he or she should have the status of an independent member," except when a want-of-confidence motion was tabled. Oke spoke at the meeting, and rather than using his speech to defend the UFO's legislative record, he provided information about redistribution, proportional representation, and the transferable vote; according to the minutes, political questions were "discussed at considerable length."[45]

All was not well for Oke. Before the nomination convention the association met twice more, and at the first meeting on 5 May some members voiced their discontent with his performance at Queen's Park, especially his position on the Adolescent School Attendance Act, which raised the school-leaving age in the province from fourteen to sixteen. Some people rose to defend Oke, and the association, perhaps for want of a viable alternative, passed a motion endorsing "the general policy of the UFO Government during the last three and a half years," despite recording concern over Oke's performance.[46]

Oke, who was not at the 5 May meeting, learned of the critical remarks and requested that a meeting be held the following week so that he could defend himself. At this second meeting Oke said that many of the criticisms levelled against him were due to misunderstandings. He described UFO legislation in detail, and then, in a rare moment of old-style politics, he reminded his audience that he had "secured assistance from the Govt for a municipal drain in Brooke Township." Oke's explanations and justifications did not satisfy everyone in attendance, however, and several questions were put to him after he finished his speech.

J.J. Morrison also spoke at the second meeting. He began his speech by denying the charge – apparently made by someone in the audience – that he was a dictator. Outlining the history of the Patrons of Industry, he argued that it had been defeated because its MPPs had been "baited by the politicians of the old parties." UFO MPPs had to avoid this trap if they wished to remain successful. He concluded with a defence of the central office's opposition to certain UFO legislation, such as civil servant superannuation, the removal of property qualifications as a condition to holding municipal office, and the Adolescent School Attendance Act. After his speech he received a vote of confidence from the association.[47]

The convention, which was held soon after, was comparatively anticlimactic. Three members, Oke, H.A. Gilroy, and Silas Smale, agreed to let their names stand as nominees. It appears, though, that Smale and Gilroy only made this commitment so that they could address the audience, since both withdrew after speaking. Smale supported Drury's policies, although he defended Oke's position on the Adolescent School Attendance Act, claiming that Oke simply did not feel that the legislation went far enough. Gilroy professed to be neutral on the broadening-out issue and "admitted that Mr Oke had a want of diplomacy which was due to a lack of education." But, he added, Oke's "greenness was wearing off." Thus Oke received the nomination unopposed, and in his acceptance speech he discussed his performance as MPP. That he sat on six committees at Queen's Park attested, he believed, to his effectiveness as a legislator. He concluded by arguing that the old parties "want to bust the UFO. I have tried for four years to educate myself and to be a credit to the party and to myself. I will fight clean to keep the old parties away from you. They are full of tricks." The convention ended with the passing of a resolution that reiterated support for the Drury government.[48]

Support for Drury did not mean that the East Lambton association uncritically accepted everything from the central office. Shortly before the election several members expressed concern about the way the *Sun* covered the UFO, prompting the following resolution: "Whereas there is a feeling among the loyal UFO electors of East Lambton that the *Farmers' Sun* is not giving adequate support to the Farmers [sic] cause during the campaign. Therefore, be it resolved that this executive ... recommend that the *Sun* devote more space and energy in combatting the many erroneous statements emanating from the Party press arrayed against us."[49]

Oke won the riding with a comfortable plurality. After the election the association met to discuss the new political reality. B.W. Fansher suggested that the UFO ought to become the Official Opposition, and that Drury should remain on as leader.[50] A Mr Auld disagreed, stating that a new leader should be chosen by elected MPPs, representatives from the central office, and riding delegates. This sparked a debate over broadening out, although exactly what was discussed was not recorded. Finally, members agreed to support the election of a leader after the December UFO convention, and they decided that East Lambton convention delegates should be instructed to recommend that a committee, consisting of members from the central office and of UFO MPPs, be struck to "consider ways and means of selecting a leader satisfactory to all and formulate plans for the calling of a convention."[51] At an association meeting later that year the assistant director, John Darville, pointed out "some weakness in the organization and voiced the opinion that the movement had not been too fortunate in the selection of leaders." It appears that Darville's comments went unchallenged, even by Oke and Fansher, who were in attendance.[52]

In the adjacent constituency of West Lambton, one sees how tensions within the UFO, often brought about by a local club refusing to defer to the central organization, profoundly affected the movement. At the West Lambton UFO Political Association's annual meeting in 1922, defeated Progressive candidate R.J. White claimed that it was wrong for party leader T.A. Crerar to pursue a broadening-out policy, which prompted a debate on the subject. Matthew White read a letter he had sent to Drury on broadening out, and UFO MPP J.M. Webster was asked his opinion on the topic. Before he could state his position, riding director Byron Young interrupted and asked Webster to explain why no democratic mechanisms had been implemented through which constituents could express their views. Young said that he had no knowledge of anyone at the local level ever being asked his or her opinion on roads policy, for example, and he asked Webster to "explain on what authority he, when speaking in the budget debate, stated that the government was carrying out the policy advocated by the UFO." Instead of offering an explanation, Webster attacked Young for not supporting the government and accused him and other members of being merely interested in seeing their names in the newspaper. Young took exception to Webster's remarks and said he "had no intention of giving servile support to any elected member." He stated that, although he felt like a nonentity after the 1919 election in terms of having input into UFO policies, at least he was going to remain true to the movement's principles, implying that Webster was not. Young concluded by restating his faith in the recall system and defiantly daring those who disagreed with him to secure the required number of votes to remove him from the position of riding director. In an effort to restore calm, W. Zealand proposed a vote of confidence in Webster, which passed.[53]

Incidents such as this serve to illustrate how complex even basic political issues could be at the local level. There were those who, although disappointed with their MPP, could express support because there was no viable alternative. The debate also clearly shows the tensions between the UFO's idealists and pragmatists.

With the call for a provincial election in 1923, Webster let it be known that he did not completely believe Drury's denial that he had tried to negotiate some sort of deal with the provincial Liberals. Webster said he would listen carefully to what Drury and the Liberal leader, Wellington Hay, had to say on the matter so that he could return to Lambton as "an honest man." Webster was again chosen to represent the UFO in the election, but, according to press accounts, there remained unresolved tensions among the riding's UFO members. Several supporters during the campaign urged rank-and-file members to put aside "petty" differences and rally behind Drury, implying that not all members supported him.[54]

GRASS-ROOTS IDEALISM

Despite the tensions within the movement, there remained UFO rank and file who displayed indefatigable energy, and who persisted in putting forward an alternative vision of society. By the mid-1920s, these members became increasingly atypical as local UFO clubs, taking the lead from the central organization and from society, began to moderate their societal critique and plan for action. But they remained nevertheless, and they persisted in attempting to fulfil the professed ideals of the movement.

F.E. Webster of Simcoe continued to criticize the old parties and the Big Interests, as well as the UFO itself. When Drury announced his intention to form a Peoples' party in 1922, much to the chagrin of J.J. Morrison, Webster dismissed press reports of the incident as nonsense. Newspaper editors, he noted in a letter to the *Sun*, would have the Ontarians believe that the movement "hinged on what Mr Drury or Mr Morrison thought." Believing that nothing could be farther from the truth, Webster recalled how the North Simcoe UFO members chose their candidate for the 1921 election: "We conducted these conventions just as we saw fit, and we neither asked nor received advice from Morrison, Burnaby, Crerar or Drury. If [Drury] ... considers we did not go far enough, or, if [Morrison] considers ... we went too far in opening our convention to all parts of our riding, why, we cannot help it, for we propose to conduct our conventions just as we see fit, irrespective of what either of these gentlemen may or may not think."

In the same edition of the *Sun* in which Webster's letter appeared, East Simcoe UFO director James Mercer argued that the Drury/Morrison split was largely irrelevant since the UFO was not a political organization. The 1919 victory was simply a protest against the behaviour of past administrations. If some farmers wished to form a Peoples' party, then they had every right to do so, but it was erroneous to claim that it was affiliated with the UFO.[55]

East Lambton UFO director Arthur E. Vance added his unique thoughts to the broadening-out debate. He accused the Big Interests of controlling the "avenues of information to the people with the apparent purpose of making sure the special privilege of their return to office." He then criticized some UFO MPPs who, he believed, had let power obscure their thinking. By adopting the old parties' tactics, they had weakened their position "instead of giving themselves strength as they would have done had they been willing to do their best and trust the people for results." Vance argued:

> It is not for any Government to say whether we shall have party government or ... group government or any other kind of government. It is for the people alone to say. As United Farmers of Ontario we have started a group movement with some considerable measure of success, and as good farmers it behoves us to plough a straight furrow and not to turn back, for we know the old ways were devious ways and led us nowhere, and we perceive that the Progressive scheme is only the old party in a new dress – a system in

which we would soon be as hopelessly lost as we were under the old parties from which we have broken away.[56]

For Vance, the problems with the new ways paled in comparison with those of the old party system.

Other farmers also voiced their opinions on broadening out. In 1921 John Houldershaw of Stayner heard that Morrison and Drury, who were reported to have been at loggerheads, were presently "cooing like a pair of doves," possibly so as to avoid alienating agrarian voters in the next federal election. Houldershaw felt that Morrison should let the people know where he stood: "If it is Mr Drury who has repented we will rejoice more than we would over one hundred UFO men who need no repentance." At a meeting in support of MPP W.I. Johnson during the 1923 campaign, William Code of Lanark had heard some people say that the acronym UFO meant "Us for Ourselves." Code disagreed, claiming that they really represented "Us for Others." He noted that the political side of the movement "was a spontaneous outburst of the people" who did not wish to elect slaves to the old parties: "The Farmer and Labor party was composed of independent thinking people whose representatives in Parliament could express themselves in a way they could not under the party system." A former Tory, Code was now an advocate for the independent policies of the UFO.[57]

These words encapsulate many of the issues that the UFO faced: the choice between forming a Peoples' party or insisting on group government; the adoption of old-party tactics by UFO legislators; central versus local control of the movement; and a press that actively sought to discredit the UFO. Identifying the problems was one thing; trying to remedy them was quite another.

Even with all these issues facing the movement, idealists continued to express their views eloquently. In 1925 "A Student of Life" in Simcoe voiced concern over criticism farmers were heaping on the *Sun* for taking an independent stand on political questions. If farmers were not prepared to support a paper that stood for the truth, then how could the organization ever prosper? The problem lay, in part, in the reluctance of some officials to tell Ontarians the truth, and the answer lay in complete candour:

Truth never yet has been harmed by open and free discussion ... Prejudice is the blind which prevents many from letting in the light of the day, while indifference and disinclination to study shut out many more from a correct understanding of public affairs ... the only hope of having an honest, representative Government at Ottawa, lies in the education of private and public opinion ... the formation of intelligent opinions will ever be a slow process so long as men put self interest or class interest before public interest.[58]

Idealistic, yes, but at the root of this commentator's words was the notion that common people were entirely capable of playing a role in how the country was governed.

In 1926, upon learning that the UFO had decided to re-enter politics,[59] F.E. Webster noted with disgust that the principle of constituency autonomy was again under attack. When the North Simcoe UFO Association originally incorporated for political purposes, the provisions of the charter were prepared under the direction of the central office. Later, when the Progressive Association was formed, the members, realizing their past mistake, clearly articulated in the charter that they were "free to act on [their] own principles." At that time Drury agreed with the clause: "But now the Political Committee of the UFO bobs up and tells us we will not be recognized by the executive unless we submit to the direction and dictates of the central office. Or, in other words, we are denied the rights of riding autonomy." Referring to the UFO's long-standing distaste for the older parties' control over their local associations, Webster wrote that "a lot depends on just who is giving the orders as to whether or not it is a crime." He also noted defiantly that the North Simcoe organization would "run our political conventions irrespective of the rules and regulations" of the central organization.[60]

In late 1925 the East Lambton UFO Political Association nominated Progressive MP B.W. Fansher as its candidate in the upcoming federal election. The UFO had shortly before decided to renew its political activity, but it had also determined that there was to be no broadening out. Defending the decision, Mrs John Darville said that it was not "a slap in the face of our town friends. There is nothing in the constitution ... that prevents sympathizers with our movement in the villages or towns voting with us [but] if you want us to hold the olive branch out to the towns ... not me. No more broadening out for this child." Also speaking at the nomination meeting was Silas Smale, who disagreed with Darville. Smale was uncomfortable with parts of the UFO resolution, particularly those that suggested that the UFO wanted nothing to do with urban people.[61]

During the 1926 federal campaign Smale wanted to "take a slam at the Liberals of East Lambton" for fielding a candidate opposite Fansher in the previous campaign: "They call themselves Liberals, and yet they transgressed every principle of Liberalism by bringing out a candidate when they knew ... there was not a ghost of a show of them winning ... For the sake of a few paltry dollars they sacrificed what they call their party principles, and they know it." According to Smale, it was that kind of "party politics we tried to get away from when the UFO was organized."[62]

The words of the rank-and-file membership were not often recorded. If those that were are in any way representative, however, then there was a considerable spirit of opposition in the movement even into the mid-1920s. At the East Lambton nomination convention for the 1925 federal election, with over three hundred people in attendance, fourteen were nominated. None but the incumbent, MP B.W. Fansher, allowed their names to stand but most of them made five-minute speeches. From these orations, one can catch glimpses of people who still carried with them the fervour of the 1919 and 1921 campaigns. Charles Stevens said that he did not aspire to office. Instead, he would "support any candidate who will legislate for all the people. I desire to see the stranglehold of the towns and villages over the rural population broken." Harvey Annett said he had worked for the interests of Canadians – not just farmers – since 1911, and that he continued to do so by supporting the UFO. John Gibson, who had been involved in the farmers' movement for forty years, believed it depended "on the farmers to bring about [their] emancipation. It is only by increased intelligence that you are throwing off the Yoke." Duncan McVicar felt that the Progressives should enter the campaign enthusiastically: "Neither the King government nor the Conservatives offer anything to us. Mr Fansher has been the ideal representative and we cannot do better than accept him."[63]

Even if these words carried little political weight by that time, they underlined the point that many farmers still believed that the system worked against them. The difference between speaking in this manner in 1919 and speaking this way in 1925 was that in 1919 such language represented a call for mass democratic action. By 1925, however, it meant nominating a candidate to fight for farmers' interests. Although the spirit of opposition is evident, there are subtle but important differences. No longer were the wishes of UFO members to be embodied in their elected officials; now Fansher was to go to Ottawa and legislate *for* them.

UFO-LABOUR DIVERGENCES

From 1919 to 1921 relations between the UFO and the ILP varied from riding to riding. After 1921 cooperation between the UFO and labour declined substantially in most constituencies for a number of reasons. First, the differences between the two movements came to the fore at Queen's Park, where Drury alienated the ILP on several occasions.[64]

Second, political support for the ILP declined among Ontario voters, leaving many of its riding associations in disarray. Third, in some cases local tensions between the two groups remained unresolved.

Even in those constituencies in which UFO MPPs enjoyed workers' support, there was no freedom from controversy. The South Lanark UFO-Labor party (as it was known) nominated MPP W.I. Johnson in 1923 to contest the riding again. In his acceptance speech, Johnson boasted of the UFO's legislative accomplishments. He pledged his loyalty to Drury and said that he was in favour of a Peoples' party. Johnson also spoke in glowing terms of his ILP colleagues who, he felt, deserved to be commended for their work in the Legislature: "They have never countenanced anything radical, never gone to extremes, but they have introduced and put through bills which made the work of the common people easier, their home life more pleasant and the hours of unemployment and sickness happier." This was a substantial reversal for Johnson who, three years earlier, when speaking in front of the Smiths Falls UFO, had referred to four ILP MPPs as "the rankest kind of radicals and reactionaries." At the same meeting Johnson assured his audience that he would "oppose any proposal ... at the next session of the Legislature with a view to the introduction of an eight-hour day" for industrial workers.[65]

In East Simcoe in 1923 UFO MPP J.B. Johnston was supported by the Orillia Trades and Labour Council. At the meeting at which this resolution of support was passed, Canadian Labor Party leader James Simpson urged workers to continue the alliance rather than align themselves with the Liberals, as they had been invited to do. To Simpson, only traitors would join up with the older parties: "The only alliance that would be recognized or allowed by the Canadian Labor Party was with the farmers. The industrial toiler should unite forces with the worker on the land."[66] The following week, however, the *Orillia Times* featured the headline "Labor Will Remain Neutral" in reference to the election. The banner was misleading; the story that followed noted that three labourers had organized a meeting, and that those in attendance decided not to help any party to select a candidate. Only after the candidates were named would labour weigh its options and possibly offer its support to one of them. The three men were part of a small group that attended a UFO meeting, from which they emerged displeased with what they heard. They then visited Conservative candidate W. Finlayson and were equally dissatisfied. Although impressed by the Liberals, they decided not to support them until a candidate was selected. One fact remained clear: this group would not support Johnston. The article was a confusing one and did not mention how many people attended the meeting. One audience member, J.D. Hean, supported Johnston, noting that the Tories never worked in the interests of workers. Responding to the charge that Premier Drury ran the province by commissions, Hean pointed out that the group that had visited all three parties was itself a commission. It is clear, however, that most in the room were anti-UFO. A Mr Childerhouse said it was strange that Johnston "should call himself a labor representative when he had repudiated labor in his presence and in the company of many others," but he provided no examples of this. Given the meeting's tone, it may have been similar to one held in 1921 that focused on discrediting the Progressives.[67]

The Liberals decided not to field a candidate, possibly because of their inability to secure labour's endorsement or, perhaps more accurately, because too many workers realized the transparent tactics employed by the Liberals to gain their support. To make matters even more confusing, the *Toronto Globe* reported that a deal had been struck between the Liberals and the UFO whereby, if no Liberal contested the seat, then no Progressive would go against Liberal candidate Manley Chew in the next federal election. In any event, the incident must have left many Labourites profoundly disenchanted since they appear to have been largely silent during the campaign.[68]

ROLE OF THE OLD ORDER IN THE MOVEMENT'S DECLINE

As seen, UFO leaders and rank and file both, in time, relinquished their alternative vision and reverted to conventional thought and behaviour. By the mid-1920s, they were motivated less, it seems, by an idealistic altruism than by a class-based self-interest. And the relationships that they had forged with the ILP, though often tenuous, eventually broke down altogether. This, apparently, was the character of the movement's decline.

With the confirmation of such a marked change in the UFO, one must turn to the issue of cause. In any examination of the decline of the UFO it is necessary to examine the role played by the society in which it operated, the society it attempted to change. Power, and the social class that had the greatest access to the wielding of power, did not let the UFO go unopposed. On the contrary, through a number of mechanisms Ontarians were exposed to messages that ultimately had a profoundly negative influence on the movement.

Some of the strategies were blatant, such as those perpetrated by the old political parties. They ranged from scare tactics to dirty tricks as these established political forces fought the UFO on every front.

The old parties may have let their guard down in 1919 and 1921, but they would not let it happen again.

Other strategies were more subtle and insidious. The agents here, the moneyed interests of the "old order," had a potent tool at their disposal – hegemony. As Bryan Palmer notes, capital is remarkably adept at presenting a world view that allows it "to bury its own interests in an avalanche of 'benevolence,' highlighting not the inequities of social relationships but their supposed reciprocities ... There are no ties that bind as effectively as those that are self-imposed, those that appear in the historical record of oppression and exploitation at the request of the very people they will secure." Equally powerful is the ability of a ruling class to develop and impose "an almost 'naturalistic' consensus, so firm in its hegemonic assumptions that the act of contestation easily seems marginalized to the point that it is out of sight and, consequently, out of mind."[69] These forces had an enormous impact on the UFO, an impact so strong that it ultimately rendered the movement harmless. Their workings were various; the old order attempted to convince UFO supporters that their concerns could be addressed through the traditional means; that UFO members were well-intentioned but misguided; or that their proposed reforms were unrealistic. But their effect was consistent and sure. The ultimate goal was consensus in the belief that the old order represented the best of all possible worlds.

Many UFO members believed that the reforms they called for necessitated change at the political level. As a result, they entered territory occupied by political parties that had considerable experience in political manipulation. In short, the UFO fought entrenched entities with vast resources at their disposal. The old parties employed an array of strategies to attempt to destabilize and defeat the movement.[70] These included patronage appointments for the local party faithful, exploiting the celebrity status of their leaders, employing scare tactics, and resorting to dirty tricks.

In order to explore the influence the older parties had on the UFO it is necessary to backtrack to the 1921 federal election, where the first significant attempts to defeat the movement are evidenced.

One way the older parties tried to sway voters was to confer rewards upon local candidates. In Lanark, Tory incumbent J.A. Stewart was appointed minister of Railways and Canals shortly before the 1921 election, thus making him a valuable asset to the riding. He was also assisted by the Liberals who, as was the case in several other ridings, declined to nominate a candidate. Indeed, influential Liberals in the manufacturing centres of Almonte, Carleton Place, and Perth declared that they would back Stewart. The reason they gave for supporting him was that, in light of the trade depression, a strong tariff (such as the one proposed by the government of Arthur Meighen) was needed to stimulate industrial growth.[71]

Another tactic employed by the old parties was to feature their prominent figures in speaking engagements. For instance, Arthur Meighen visited Simcoe during the 1921 campaign. Speaking in front of two full halls in Orillia, the prime minister defended his government's record and pointed out the deficiencies of the other parties. He bombarded his audience with an array of statistics that, in his view, attested to Canada's growing economy, and he boasted that no breath of scandal had touched his government. Making certain to discredit his opponents, he hinted that there was an "arrangement" between Liberal leader Mackenzie King and Progressive leader T.A. Crerar such that certain constituencies were straight two-party battles between either Tories and Progressives or Tories and Liberals.[72] Pronouncements of this sort would have been given serious consideration – even in the 1920s party leaders had celebrity status – and the noteworthiness of having the prime minister in one's riding undoubtedly meant that his message had an influential effect on voters.

The old parties relied on negative messages as well, primarily those that discredited their opponents. In 1923 East Simcoe Conservative candidate William Finlayson brought in Tory MP W.F. Nickle, who claimed that Drury had said that, if the UFO did not win enough seats to form the government in the election, then he would consult with elected and defeated UFO candidates and the UFO executive to determine a course of action. To Nickle, this was hopelessly undemocratic: "Elected members should be the spokesmen of the people ... if the people of East Simcoe elected [UFO candidate] Mr. Johnston ... [he] could not tell now where he will be or what he will do." J.L. Hartt of Orillia also spoke on Finlayson's behalf at the meeting and asked the audience to remember the image of Sir James Whitney riding to Queen's Park on his bicycle, and of his simply furnished office, in contrast to the extravagance of members of Drury's cabinet, with their chauffeured American cars and luxurious offices.[73]

In East Lambton during the 1925 federal election prominent Liberal I. Greenizen stated that he supported paying bounties to oil producers in the area in lieu of protection. He then claimed that his Progressive opponent, B.W. Fansher, was also on record as supporting these payments, although no evidence to this effect was provided. Even so, Fansher voted against them in the House of Commons. For Greenizen, this represented blatant hypocrisy.[74]

Another strategy employed by the old parties was to try to scare voters back into the fold through the propagation of misinformation and malicious distortion. In January 1921, well before any federal election was called, Joseph E. Armstrong, Tory MP for East Lambton, held what amounted to campaign meetings in which he warned that, if the Progressive platform of free trade was implemented, it "would wipe out every industry" in Forest. Thanks to the tariff, he claimed, over two hundred American industries had established operations in Canada in 1920 alone. Once the tariff was eliminated, they would disappear.[75] In Lanark, Tory supporters alleged that the Progressives were pure free traders, and that their policies would lead either to the ruination of domestic industry or, worse, to annexation with the US.[76]

During the 1921 campaign stories began circulating in the press that Prime Minister Meighen had referred to the Progressives as dangerous bolshevists. In response to the allegation, Meighen issued a clarification:

> What I said … was that, though they themselves were wholly opposed to Bolshevistic principles, they had trailing behind them and allied with them men of seditious principles in this country, and that statement was true then and is true today … Did you follow the … Provincial contest in Alberta and see that Mr William Irvine and Mr Woodsworth and some others were taking part on behalf of the United Farmers, and did you know that one of them was just through serving a term in the penitentiary convicted … of sedition?[77]

By the time the clarification was issued, though, the damage was done. Even the retraction planted doubts as to the competency of the UFO. After all, what was to be made of a movement that could not prevent dangerous bolshevists from infiltrating its ranks?

There is considerable evidence that the old parties occasionally engaged in dirty tricks. For instance, South Lanark MPP W.I. Johnson spoke at the annual meeting of the county UFO about a "certain gentleman" who had been speaking out against alleged extravagances and scandals at Queen's Park, and of a "certain sheet attacking the Drury Government [that] had been mailed" to households throughout the province. Johnson asked, "Where did the money come from to pay for that sheet?" For him, the answer was clear; the Big Interests had financed it in order to generate support for the old parties. Dan Hogan, the UFO riding director for Lanark, told supporters that the Conservatives had booked every hall in Perth "so that the Progressives would not be able to hold any meetings" there. In North Simcoe it was alleged in the local "UFO Column" that Colonel Currie or his supporters had sent a leaflet, "without signature, through the mails to the electors of North Simcoe containing lying, slanderous statements" about the Grain Growers Grain Company.[78]

Some attempts to discredit the UFO backfired. During a 1923 campaign meeting North Lanark Conservative candidate Thomas A. Thompson, accompanied by Kingston Tory MPP W.F. Nickle, said he sympathized with farmers' complaints, but he believed that it was not in Ontario's best interest "to have one group governing the country." He was suspicious of anyone who pitted class against class and "creed against creed." Thompson then attacked the UFO's extravagance and claimed that civil servants had been forced from the Legislature building so that cabinet ministers could make room for themselves and their wives. At that point, according to a press account, Thompson "turned to Mr Nickle for corroboration of the statement, but the latter shook his head non-committally."[79]

The old parties also relied on the power of slick but superficial presentation, an effective strategy whereby they appeared to understand why farmers were angry while calmly assuring them that forming a mass democratic movement was not the best means to accomplish change. This meant utilizing balanced tones and seemingly logical arguments. Manley Chew, the original choice of Labour in East Simcoe who was rejected by the UFO and then subsequently accepted the Liberal nomination, was a master of balanced, almost seductive tones. He ran a large advertisement in an Orillia newspaper with the headline "If You Are Thinking of Voting Farmer." In it, Chew acknowledged that his Progressive opponent, Thomas Swindle, was well liked: "But is he really the best man to represent East Simcoe in Parliament? Would not an experienced man be a better representative? And is it not true, if the farmer candidate were not running, all the rural vote would go to Mr Chew?" Chew also claimed that he was the farmers' true and trusted friend, and that he would "sit on the Government side if either Mr King or Mr Crerar is in power." Chew, "the People's Candidate," expressed no commitment to cleaning up government or to allowing greater citizen participation.[80] Nonetheless, he claimed, farmers would find themselves in a win-win situation if he were elected.

Chew benefited from support from the Liberal *Orillia Times*. Its editor thought it "regrettable" that there were three candidates in East Simcoe since it split the anti-Tory vote between Chew and Swindle. He noted that in many rural ridings the Liberals had not nominated a candidate, so as to make the going easier for the Progressives. As East Simcoe was largely urban, "the Farmers should have given way and helped to achieve their own ends" by supporting Chew. The Tory candidate would then be easily defeated. "Will they do it, or will many of those in East Simcoe stand pat, vote for a *mythical* principle, and perhaps by their mistaken action elect the candidate and support the policy they have been denouncing so vigorously?"[81]

In other instances, Conservative adherents relied on seemingly rational arguments as they extolled the virtues of the tariff. J.A. Stewart informed Lanark farmers that the US Fordney-McCumber Tariff had drastically reduced Canadian exports to American markets. Consequently, it was in the farmers' best interest to support the Canadian tariff walls, since the farmer now had to "find a market for the products heretofore sent to the United States – he is now more than ever dependent upon the home market. The more employment there is in our towns and cities, the larger market there will be for the farmer's produce. Crerar's Free Trade would diminish employment in Canada and thereby shrivel up the home market of the Canadian farmer."[82] Stewart's reasoning was, at best, shaky. First, as with many old party candidates, there was the tendency to misrepresent Crerar as a pure free trader. Second, in his arguments he overlooked the potentially lucrative European market that policy makers often claimed should be exploited more fully. Finally, Stewart failed to acknowledge that the Fordney-McCumber Tariff might itself have been at least partly retaliatory, and that, as it aptly demonstrated, problems were created as much as they were solved when tariff walls were erected.

Other strategies of the old parties included the appropriation of UFO positions and the exploitation of the emotion connected with patriotism and the memory of fallen friends and family. The pronouncements of J.A. Stewart are a clear illustration of the former strategy. In his election advertisements, he reiterated Conservative policies and then added a few touches of his own. He promised, for example, better "co-operation between employers and employed," improved marketing for agricultural produce, extended rural credit, and improved social conditions for those in rural areas.[83] In his campaign tactics, North Simcoe Tory MP Col John A. Currie exemplifies the second strategy. In many of his advertisements ample mention was made of his military background. Some of his promotional pieces, for instance, featured a photograph of him in uniform (medals, of course, prominently displayed) with the words "Lest We Forget" printed underneath it.[84]

If these strategies proved less than effective, then the old parties resorted to the ploy of unravelling the farmers' confidence in their ability to make straightforward, positive choices. For instance, in a federal by-election in Lanark in 1922, once it became known that there would be no Progressive candidate, both Liberal and Tory forces went to work to try to win the UFO vote. The Liberals assured members that "Liberals and farmers are not far apart on most questions ... If you let the Conservatives win you strengthen the Opposition to your interests at Ottawa." Even so, the Liberals had to reach fairly far to find an example of how King had aided farmers. A subsequent advertisement cited, as its sole example, King's actions with respect to the Crow's Nest Pass Agreement: "You may say that this does not affect you here in Lanark. But you have sons, daughters or other relatives and friends in the West in whom you are interested ... the Liberal Government's work in helping the Western grower shows that the Liberal Party has the farmers' interests at heart and can be counted upon to assist Agriculture in Lanark County and every other county."[85]

The Tories stood the issue on its head and asked farmers to consider what would happen if a Liberal was elected: "Lanark ... would make the King Government independent of the Progressive support in the House. Progressives know they could not get any legislation enacted in the Progressive interests. Is it not, therefore, to the very best interests of the Progressives and UFO to vote for [Conservative candidate] Dr Preston ... and maintain the whiphand the ... Progressive party has today over the King Government?" It must have been demoralizing for UFO supporters to realize that their vote could be used only for strategic purposes, rather than for the direct support of a candidate who upheld their principles.[86]

The cynicism that pervaded politics must also have been demoralizing for UFO members and their leaders. During the 1923 campaign East Simcoe Conservative candidate William Finlayson presented his view of how things should be: "There had not in the past been any great difference whether the Liberals or Conservatives were in power at Toronto. When one got stale, that party was put out and the other elected to office. This was the time for Liberals to vote for the Conservative candidate, so as to help defeat the Drury government. It was not right that any occupational group should hold office. The duty of all was to put an end to the unbusinesslike government at Queen's Park."[87] An analysis of these remarks could be a book in itself. What is noteworthy here is the frankness of Finlayson's admission that the two older parties were virtually identical. Unlike the UFO supporters who agreed with this analysis, though, Finlayson saw it as a positive feature of the Canadian political system.

Another strategy of the old parties was to defuse the UFO's radical dynamic by charging that it was merely composed of disguised Liberals. Although many UFO members vehemently denied that there were any formal, organizational links or interconnections with respect to philosophy and platform, there were critics from both old parties who argued to the contrary. Whatever the merits of this allegation, the fact remains that it was used frequently by the Liberals as a compendium for other seemingly rational and convincing arguments. The rhetoric was indeed compelling in its clarity and assuredness. After Drury was nominated by the Progressives in North Simcoe in 1925, for instance, the Liberals of the riding decided not to field a candidate because, quite simply, a party with fiscal policies "similar to that of the Liberal party" was already contesting the seat. Thus, all Liberals were urged "to make every effort in opposition to the Conservative party and its policy in this riding."[88]

The reality of the situation was that Liberals could not afford to alienate farmers to the extent that the Tories had done. As a result, their oppositional strategies were somewhat more subtle, though they were equally effective. During the 1925 campaign, for instance, West Lambton independent Liberal candidate W.T. Goodison (who also claimed to speak for Progressive supporters) promised to make decisions based on the wishes of his constituents.

Goodison offered a comfortable choice – he often spoke of the need to pursue a middle course, thus convincing many farmers that he would address at least some of their concerns. During the campaign, for example, he proposed a compromise regarding the tariff. With respect to Senate reform, he promised that something would be done to bring it "in closer touch with the people of the country."[89] But there is also something subversive at work here. Whether it was intentional or not, he and others like him were, as Palmer put it, burying UFO interests in "an avalanche of 'benevolence'" that highlighted "supposed reciprocities" rather than actual inequities. Ultimately, they were attempting to develop and impose "an almost 'naturalistic' consensus.[90]"

It is possible to go on at length about the kinds of "naturalistic" assumptions Ontarians were exposed to. However, for a succinct illustration of the kind of opposition UFO members faced after their successes at the polls, I shall turn to the efforts of a Simcoe newspaper, the *Orillia Times*, to "educate" its readers. In the wake of the populist upsurge, the *Times* ran a series of editorials on two contemporary issues: group government and proportional representation. At the outset group government was virtually dismissed as a viable alternative since, it was noted, even the farmers were divided on the subject; in any case, it was debatable whether the group model could provide better government than the traditional two-party system.[91]

In subsequent editorials the paper continued the attack with a detailed accounting of the concept's weaknesses. First, group government led inevitably to coalitions, which meant that there was the danger that a minority group might "dominate the situation." In fact, it was quite possible "for a small minority group of insistent agitators, holding the balance of power, to foist upon Parliament and the country legislation the people do not want. It would introduce into our political system bargaining and compromise," although it was not mentioned why compromise was such a dreadful concept. The group system would also produce a government that was neither "stable nor trustworthy," and "practically devoid of responsibility."

The two-party model, it was argued, was the result of over two hundred years of evolution, and it had "proven suitable to the disposition and temperament of the Anglo Saxon people." In addition, the old system made for "stable and firm government," which was exactly what the country needed: "Whatever may be said of Group Government as a theory, this certainly is no time to experiment with political, social or economic issues."[92]

The editor noted that, under the group system, legislators would be elected by proportional representation, of which the *Times* was even more critical than it had been of group government. After admitting that the existing system sometimes produced unfair results, it protested vigorously that it was still a far better system than proportional representation, which, if implemented, might result in a small group with the balance of power dictating government policy, which in turn would make for laws that most people did not support: "We should never forget that the very essence of our liberty is contained in the fact that a Government must have a majority in our Parliament." Proportional representation would also bring enough candidates into the field to divide the vote, which would lead inevitably to the election of "representatives of minorities who have not behind them popular approval for their fads and fancies." Other arguments against proportional representation were that vote counting would be slow and tedious; that a British royal commission recommended in 1910 against adopting it; that the system did not always produce an accurate reflection of votes cast to seats earned; and that the "cost of elections to the candidate over a large constituency eliminate[d] everyone but the rich candidates who can travel the riding and spend plenty of money." Having demolished proportional representation as a legitimate alternative, the author of these *Times* editorials then identified the important issue at hand – the need to institute a more equitable distribution of seats according to population, since some rural ridings, such as Simcoe East with fewer than fifteen thousand voters, elected members on an equal basis with much bigger voting populations.

In its justification of the existing electoral system, the *Orillia Times* advanced a few positive arguments. Implicit in the articles was the notion that the existing system represented the culmination of centuries of British constitutional development.[93] It was inherently balanced and inclusive since the use of small constituencies and of two parties tended to "eliminate race, creed and class candidates." In addition, there was a measure of stability in constantly electing majority governments: "At least there is always safety in majority rule." Finally, under the first-past-the-post system, the process was not only solid and well grounded but also easy to understand: "No other system yet devised is an intelligible or speedy."[94]

It is here that one discerns hegemonic forces at work. In effect, the writer of the editorial pieces was asking UFO members who detested racism to bolster a racist system; who became politicized because they felt themselves to be different (and to be exploited as different) to support a system that could not be disrupted by minorities; who were young and dynamic and eager for change to uphold the values of tradition and stability. The terms of the debate, however, were not phrased in that manner. As I have demonstrated, the persuasion to which UFO members were subjected was much more subtle than that.

Although some editors, such as *Almonte Gazette* editor James Muir,[95] were friendly to the UFO, the majority sided with the arguments put forward by the *Times* editor.[96] In an age when so-called political experts were sought out by rural weeklies, these were powerful ideas.[97]

The effects of these forces can be seen in the decline of the political activity of the movement. After the mid-1920s the political side of the UFO was largely inactive. In Lambton, Progressive Bert Fansher recaptured East Lambton in the 1926 federal contest, but by then he was running on a low tariff, low tax platform; many of the features (and much of the enthusiasm) of the previous campaigns had disappeared.[98] Provincially, L.W. Oke retained his seat in the 1926 election and lost in the 1929 contest. By then, however, Oke was better known for his concern over the financial interests of East Lambton farmers than over challenging the Big Interests.[99] In Simcoe, aside from Drury's candidature in the 1925, 1926, and 1930 contests, no Progressive candidate ran for a federal seat after 1925.[100] Provincially, T.E. Ross ran and lost as a Liberal-Progressive candidate for East Simcoe in the 1926 election, as did John Mitchell in Southwest Simcoe, thus ending direct participation in provincial politics. In Lanark, UFO political activity effectively ceased after the 1926 federal election, when a Liberal-Progressive G.W. Buchanan contested the seat and lost.[101] It was the last time a UFO candidate ran in an election in Lanark.[102]

The disillusionment of UFO supporters can be seen even before the post-1925 elections. In 1919 the provincial voter turnout was 72.6 percent. In that election, many of the young UFO supporters may have voted for the first time. Certainly all of the women voters did. In the 1923 provincial contest the voter turnout was 54.7 percent. This represented not only a dramatic drop in voter turnout but also a record: 54.7 is the lowest percentage of voter turnout *ever* in an Ontario election.[103] Although one can only speculate why this dramatic decline in voter turnout occurred, it is reasonable to assume that it partly reflected apathy and disillusionment in those who supported the aims and ideals of the UFO.

At the provincial level, in 1932 the UFO briefly affiliated with the newly formed Co-operative Commonwealth Federation, and then endorsed H.H. Stevens's Reconstruction Party in the 1935 federal election. That same year the Ontario Federation of Agriculture was formed as part of the Canadian Chamber of Agriculture (later the Canadian Federation of Agriculture). The UFO, for all intents and purposes, became the educational wing of the federation, concentrating mainly on encouraging orderly marketing of farm produce, until it disbanded in 1943.[104]

CONCLUSION

The elections of 1919 and 1921 filled UFO members with exhilaration. Hitherto unpoliticized farmers throughout Ontario became aware of the systemic inequities in the society in which they lived, and they pursued these matters through individual study and, more importantly, through collective action. Virtually all aspects of the province's social fabric were held up for examination and in many cases they were found wanting. In this exciting environment members sought to advance alternatives that seemed to them far much more equitable, democratic, and just. By contesting elections, a considerable measure of political power was achieved. It was natural to assume, therefore, that change was on its way.

The exhilaration faded, however, as members found that implementing change was going to be far more difficult than they had first envisioned. The movement's leaders seemed reluctant to work towards meaningful change; instead, they asserted that what was needed was balance, compromise, and moderation. In addition, there was a lack of consensus among the leaders on matters as fundamental as how the UFO should develop. The lack of zeal and squabbles undoubtedly trickled down into the rank and file and, in some cases, resulted in local groups merely debating issues rather than trying to devise ways to attain specific goals.

MPPs and MPs proved to be of little assistance. These individuals employed fiery rhetoric during their inaugural campaigns and, quite probably because they did so, were elected. Once at Queen's Park or Parliament Hill, however, with few exceptions these legislators began sounding and acting like the old-style politicians they had defeated. When challenged by the membership about their inaction, they reverted to traditional behaviour, choosing to rely on patronage to garner support. For many, it did not work.

Moreover, relations between the UFO and the ILP, always tentative, became more distant by the mid-1920s. Each group had a specific agenda, and each became increasingly unwilling to compromise or to join forces to fight the common foe, the Big Interests. The inability to forge a lasting alliance had profound effects on both movements.

As if the UFO did not have enough internal problems to contend with, there was another foe – hegemony. Prevailing ideas, attitudes, and mores filtered into the movement, sometimes so subtly that they were not perceived. Consequently, many rank-and-file members abandoned their alternative vision for the possessive individualism that pervaded the society in which they lived. Pinpointing this shift is impossible; it occurred slowly and without a significant event to account for the change. But the evidence strongly suggests that it happened all the same.

Thus far, only the traditional political activity of the UFO has been examined. I now turn to two important components of the movement – the United Farm Women of Ontario and the UFO's cooperative enterprises – to explore how, in a less-politicized context, UFO members were briefly able to free themselves from the prevailing worldview but then ultimately lost their ability to see things differently.

5 "Citizens Instead of Wards": The United Farm Women of Ontario in Lambton, Simcoe, and Lanark Counties

> Feminism is woman's "restless desire" for deliverance from the false position of subordination in which she has been placed by the reactionary forces in the civilization and progress of humanity of the middle and the dark ages, and the irrepressible determination to resume the position of just honor and esteem in which she was anciently held.
>
> Emma Griesbach[1]

Shortly after the formation of the United Farm Women of Ontario (UFWO) in June 1918, Emma Griesbach wrote an optimistic article celebrating the recent awakening of farm women throughout the province. There was considerable enthusiasm for the new organization, and she believed it would be the ideal vehicle by which women could attain the equality they had long deserved. Griesbach also observed that it was up to the men in the UFO "to give the sisters to understand beyond any doubt or question that the presence of [women] in the clubs is desired and will be appreciated."[2] This, she believed, would be the true test of the UFO's commitment to equality for all, and it would ensure the continuing success of the UFWO.

The United Farm Women of Ontario has been largely ignored by historians. Moreover, the few short studies of the organization are dismissive about its having had any long-term influence.[3] According to Pauline Rankin, for example, the UFWO represented Ontario farm women's "submerging their commitment to women's issues and rallying with their husbands and fathers, a strategy that eventually contributed to their limited impact and subsequent demise."[4] At first glance this may seem to be so, but upon closer examination does Rankin's conclusion stand up? Were there not occasions when the UFWO enabled women to push the boundaries within which they found themselves? And did these women learn anything from their experience in the movement?

Most studies of women in organizations established and dominated by men take the position that the women, having allied themselves with male-dominated organizations, were working within a patriarchal framework that profoundly limited opportunities for fundamental change.[5] The extent to which women were able to advance their own agenda within this framework usually stands as the measure of their success. In this chapter I argue that, although the UFWO offered women hitherto unheard-of opportunities to explore and expand upon issues that they felt were important, and although they were able to effect some reforms within the farmers' movement that were to their benefit, many of their initiatives came to naught. At times this was because of UFO inactivity on women's issues; at other times it was because of general societal constraints on women. Given recent scholarship in the field, these conclusions are entirely obvious. But the experience of women in the UFWO amounts to something more than a failed outcome. Thus, I also argue, in line with this study's larger thesis, that women who struggled to influence the farmers' movement must be viewed, first and foremost, as historical agents who attempted to break free from the prevailing orthodoxy and play a more meaningful role in how their lives were shaped.

In this chapter, close attention is paid to letters written to Emma Griesbach – known as "Sister Diana" to readers of her weekly page in the *Sun* – from women in the three counties under consideration.[6] These letters provide valuable insight into the concerns of rural women. That said, although the value of such letters as historical source material is indisputable, they are not always available to researchers. For instance, the UFWO was as significant a force in Lanark County as it was in other parts of the province. Countywide membership rose until at least 1926, when most other UFWO clubs were experiencing a severe falling-off in terms of numbers. Some UFWO clubs in Lanark survived even into the early 1940s.[7] Nonetheless, the voices of Lanark women were rarely heard in the pages of the *Sun*. Those who did write to the paper often used pseudonyms rather than their real names, and their letters were frank in subject matter and in the solutions they proposed to the problems that farm women perceived.[8] Unfortunately, the use of pseudonyms and the correspondents' frequent failure to mention where they were writing from ensure that many relevant letters have not been included here.

THE CONCERNS OF UFWO MEMBERS

Rural versus Urban Society

The differences between urban and rural society were felt as acutely by women as by men in rural Ontario.[9] Farm women were just as offended as farm men at being perceived as uncultured hayseeds, and they resented the fact that urban women enjoyed the benefits of labour-saving technology to which they themselves had no access. They chafed at the isolation of their lives, aware of the social activities that would be available to them in cities and town.

The tension between farm and city was exacerbated by the onset of the First World War. Writing to the *Orillia Packet*, "E.A." pointed out that, if urban people helped out more and talked less about farmers' not going off to fight, then something might actually get accomplished. The writer further observed that women in towns and cities were still able to purchase new clothes, while farm women were "as a rule dressing plainer and working harder" than ever before.[10]

The subject of work – or, more properly, the disparities in workloads – became a common theme in the expression of rural/urban tension. In a letter from "A.A.G." of Simcoe to Sister Diana was this description of the work she undertook on the farm: "I have four small children, do all my own work, sewing, mending and all ... I milk, feed calves, pigs, hens, etc. When haying and harvest comes I work in the field day in and day out and nurse a baby besides, and I get no days off. In ten years I have been to one picnic." The final sentence in this passage touched upon the isolation that was felt so acutely by farm women. The loneliness of farm life was a problem identified by many farm women, and for some the remedy rested with the UFWO and its emphasis on meeting and addressing issues as a group, not as isolated and distant individuals.[11]

In some cases UFWO members blamed the women themselves for their sorry state. "Sister Lou" wrote to Griesbach that she was "sick to hear farm women grouch ... about how hard they work and keep at the same old gait about a thousand years behind the times in their housework and never make the least effort to get out of their rut." The writer may have been in a better position to than most farm women alter her routine; some farm women complained that they were unable to enjoy hobbies to the extent that city women did because they were too busy to do so.[12]

To return to the subject of workload disparities, "Sister Blanche" of Lambton, a widow, felt that during the Great War her efforts were needed more in the field than in the house. As a result, she took the government's advice and hired a town-woman to keep house. In order to make the work easier for the hired help, Blanche had her eighteen-year-old daughter do all the washing and churning. After two weeks, however, the town-woman resigned, claiming that there was too much work for her to do. Blanche then compared this woman's duties to those of a farm woman she knew who was also a widow with one son registered for the war: "She feeds thirteen head of cattle, twenty-three pigs, three cows, milks eighty quarts of milk daily, feeds and turns that milk through the separator alone, cleans all the stables and pig pens, cuts mostly all her own wood, manages four of a family, feeds and clothes them on less than six hundred dollars a year." Not a pleasant picture, made even more unpleasant for Blanche when she thought of city people who were going out and enjoying themselves "while our soldier boys perish on the bloody battle field."[13]

Some women spoke with a mixture of pride and resentment when describing the amount of work they did on the farm.[14] In a debate held by the Smiths Falls UFO and UFWO on the resolution that the farmer's wife and daughter were superior to the farmer and his son on the farm, Miss Minnie Armstrong contended that women looked after the vegetable garden and the poultry, operated a free laundry and barbershop, and were often responsible for much of the repair work on the farm. Mrs Russell McDonald added that labour-saving devices had made work less strenuous for men, but farm women were not able to acquire similar devices of their own. She also knew one woman who milked twenty cows daily "because her husband had never learned to milk." Mrs R. Bowen argued further that men often employed hired help, a luxury often unavailable to farm women.[15]

Tensions between urban and rural women continued after the First World War. In a letter to the *Orillia Packet* in 1920, "A Farmer's Wife" pointed out that in the past hotels had provided farmers and their wives and children with a comfortable meeting and resting space. At the end of the day, when the shopping was done, the family could wait in the hotel parlour. Parcels could even be delivered there. The situation, however, had changed: "Now you see a notice that the accommodation of the hotel is only for guests. So it is no more a public but a private house. Farmers' wives and children wait about in stores, or round the doors." It was no wonder to her that many farm women used mail-order houses for much of their consumer goods. The writer also noted that ordering by mail made more sense during winter, because drives to town in the cold could then be avoided. She also mentioned that there were no more tying posts for horses in Orillia, which caused problems for her sixteen-year-old daughter who enjoyed riding into town. Her complaint regarding hotels corresponds with the call for the construction of restrooms in towns for the use of rural women. Rejecting the reliance on hotels for such comforts, Griesbach wrote that "women never feel quite so much at home at a rest room in a private building. It gives one a feeling of accepting charity; but in the room provided by the public she is as much at home as anybody, and even has a small sense of ownership."[16]

Alice Webster of Creemore hoped that rural women were actively supporting the Drury government's efforts to electrify farms. The "bellowings" of Orillia's mayor that hydro should be developed for use in cities "suggests a scene we often witness on the farm when the biggest, fattest hog in the pen lies full length in the trough to keep the others out." She went on to suggest that a prod with a pitchfork would be the "most considerate attention such animals deserve."[17]

Louise Collins of Stayner supported Drury's broadening-out plan if for no other reason than that it might enable rural and urban people to understand one another better. Even so, Collins could not escape the feeling that farmers were characterized as hayseeds by urban people. She pointed to *Toronto Globe* articles that criticized farmers for not understanding how the doors operated on the new Toronto streetcars. "How many times," she asked, "have I gone to the pump ... to show the stranger from the city how to get water when he didn't understand? When is the press and the crude city people going to stop such slandering, and wake up to the fact that the Progressive farmers do not envy anything in the city." All Collins wanted for herself and her fellow farm women was "a square deal."[18]

The Press and Its Distortions

As with members of the UFO, many UFWO members believed that the press – especially the urban press, from which many rural weeklies reprinted items – often distorted facts to serve the agenda of the Big Interests, and they recognized how this was done. Griesbach noted that the Big Interests drove ideas into people's heads "by means of printer's ink ... they iterate and reiterate; they publish it in every manner, and they keep on publishing it until they believe they have achieved their end."[19] Simcoe's "A.W." believed that those who criticized farmers from what they read in newspapers "should qualify their criticism by adding 'according to press reports,'" because the articles were often inaccurate. The correspondent was likely Alice Webster, who asked in subsequent correspondence whether women ever saw "in the society news of the city dailies [anything] but descriptions of how the women and the reception rooms were decorated and never a line to indicate that the women did anything more than display their finery like dummies in a shop window."[20]

Griesbach also alluded to the Big Interests subsidizing newspapers in Canada "for the sake of getting their propaganda over to the people." The sort of propaganda described by Griesbach, however, did not necessarily apply solely to newspapers. Alice Webster supposed that other sisters had grown weary "of finding out of their mail the

'Made in Canada' literature which insults our intelligence by asking us to put in manufacturers' pockets what should go into the public treasury and then call it patriotism!"[21]

Inaccurate and dishonest reporting often increased with election campaigns. During the 1923 Ontario election, a woman who identified herself as "Centre Simcoe" argued that "the city dailies are so anxious to sow distrust in the minds of the farmers, and some farmers seem content to let them."[22] Again, readers had to be very careful when reading "objective" journalism.

The distortions and outright lying of the city press disgusted many farm women and may well have reinforced their resentment of urban society. The view that the urban press was serving a dubious agenda may also have made women more sceptical when assessing political and economic questions, both of which, as will be seen, were perceived to be dominated by urban Big Interests as well.

Militarism

Militarism or, more precisely, anti-militarism was a theme frequently addressed by Griesbach. Unlike many peace activists, who either muted their message or reversed their opinions during the war, she continued her anti-war articles for the *Sun*.[23] She often warned her readers not to be led "into thinking that the chief purpose of universal compulsory military training is the defence of Canada." According to Griesbach, the generals who advocated such a policy had motives of a more base character, such as big salaries, easy positions, fine clothes, overseas travel, private cars, decorations, and automatic promotions. Most importantly, in her view, they saw "power and authority, which their souls love." As to the causes of war, Griesbach saw virtually all wars arising out of one or more of the following motives: love of battle, love of adventure, love of plunder, love of power, and the love of conquest and territorial extension. In one of her articles on the topic, she gave a thumbnail sketch of the history of wars involving Britain. She pointed out that not one of the wars she described "arose from the wish of the people, or by the will of the people, or for the good of the people."[24] Those who believed that the Great War had been waged to "make the world safe for democracy" or that it was a "war to end war" overlooked the fact that the war had enriched the wealthy and provided more misery for the poor. Griesbach also chastized those who spoke about a period of reconstruction. To her, talk about such lofty goals was empty prating if people refused to acknowledge "that in human society the first constructive principle is a real, universal, cogent and ever-active sense of human brotherhood."[25]

Alice Webster of Creemore addressed other aspects of warfare in one of her letters to Diana. Referring to a farm family with whom she was acquainted, she remembered how the last remaining son on the farm had been conscripted, "even though twenty thousand trained Canadian soldiers in England had nothing to do but rehearse for moving pictures." Yet farmers were criticized for complaining about the war and for allegedly wasting crops during the conflict, while those who donated money and nothing else to the cause were lauded as "patriots."[26]

"Seabird" of Simcoe maintained that the vast sums of money spent on military preparedness meant that money was effectively stolen from the people. "Fidelia" of Simcoe blamed religious institutions for not taking a stand against the war. Prior to the outbreak of hostilities, she argued, it was easy for the churches to preach humanity and peace. Immediately after war broke out, however, they fell in right behind the state in its war effort: "We see the outcome ... instead of the 'Church' acting in the capacity of a peace-maker in all belligerent countries, calming men's passions and moderating their fury ... we have the clergy advocating the very forces Christ was continually deprecating, self assertion, temporal power, hate and enmity!"[27] All of which went to prove to Fidelia that there was no real separation between Church and state. A few months before Fidelia made these remarks, a woman who signed herself "Lambton County Farmer's Wife" wrote to the *Sun* with suggestions on how to stop the war. The wording of the letter was such that the editor feared "we might be committing an offence against the censorship regulations" if it were published.[28]

Democracy

The attainment of a meaningful democratic form of government was a central concern for many farm women, and it is their musings on the topic that reveal (albeit unconsciously) many elements of anarchism. For many UFWO members democracy meant much more than casting a ballot every few years; it meant encouraging people to accept the responsibilities of citizenship and continuously contributing to public policy formulation.

The UFWO leadership contributed in part to this concern. Margery Mills argued that the UFO was a model of democracy because its policies derived from resolutions passed at various farmers' meetings. Each point was discussed clause by clause, and was then accepted in open convention. Mills could not think of any political group in which policy was made in a similar manner. She then gave a brief lesson on how democratic forms would come about in government:

When a candidate, instead of following the old-established order of procedure and submitting to the electors a policy which some group of men other than themselves or himself has formulated, for their endorsation, accepts a policy formulated by the people ... he changes the whole system of government – makes it democratic. The responsibility of the government belongs to the people ... and their representative may, and should, shift the responsibility from his shoulders to theirs.[29]

Obviously, Mills hoped that the UFO government would initiate this process.

Griesbach also wrote extensively about democracy and had many insights on the topic for her readers. One of her paramount concerns was the idea of self-government as a precondition for the emergence of democratic forms. For her, democracy could never be attained until "the units of a community are self-governed in their individual lives, and when they have become sufficiently enlightened to see that the good of each is bound up in the good of all and that no one can grasp more than his due without disastrous consequences to himself and to others."[30] Decentralized small-unit democracy was the key for Griesbach, as it was for many members of the UFWO.

The granting of the vote to women unleashed a flurry of comments from UFO supporters. Shortly after receiving the federal franchise, some farm women heaped derision upon those who voted for the party rather than the candidate in the 1917 federal election. "A Mad, Bad Sister" from Simcoe County wrote to the *Sun* claiming that she would vote for the devil himself if she thought he would advance policy that she approved. Even if Parliament were "a congregation of Saints," she asked, "what benefit would the people have ... if they put through stupid, blundering, ill-considered measures" such as the War Time Election Act?[31]

Cooperation

Cooperation was another topic of interest among farm women. Griesbach often wrote about cooperatives, arguing that they were one of the few tools farmers could use against the combines of the Big Interests. Some of the ventures farm women explored were designed to obtain better prices for their produce. Others were directed towards saving money and improving conditions for farm women. Griesbach campaigned for the establishment of cooperative laundries, creameries, bakeries, electrical plants, and telephone systems, to name but a few examples. She often told her readers how cheaply a community laundry or a cooperative kitchen could be established, and what benefits it would bring to the community.[32] Alice Webster agreed with Griesbach on the need for cooperatives to address women's needs, and she argued forcefully that co-operation should be the goal of all human activity. Although there were many obstacles to the achievement of this goal, she concluded that "the only insurmountable obstacle lies within ourselves."[33] In her talk before Collingwood's Union UFWO, "W.H.B." of Simcoe mentioned the need for recreation and community spirit. For her community spirit meant "cooperation, doing things together for one another; it means real neighbourliness in other words, brotherly love as the Apostle Paul puts it."[34] As this and other references indicate, farm women eagerly embraced the cooperative ethos of the United Farmers' movement and sought to apply it to many aspects of their lives. The fate of the movement's cooperative ventures is addressed in the next chapter.

Relations between Farm Men and Farm Women

Other than in idealized accounts of what farm life should be like, UFWO members made only scant reference to the relations between men and women in the farmers' movement. Nonetheless, some insight into these relations can be found in comments, written or spoken, by the UFWO élite.

In the early years of the UFWO some members expressed hope that the organization would elevate farm women to a position equal to that of their male counterparts in both public and private affairs. At a UFO picnic in Simcoe in 1920, Griesbach noted that women were equal to men in the movement:

> We are governed by the same constitution. Just as in the home one will take one department and another ... it is much the same in the farm organization and at the same time, just as in the home, we all give each other support and in time of need, a helping hand. We United Farm Women realize what an immensely important ... part public business has in the happiness and welfare of every man, woman, and child in the Dominion, and as intelligent, conscientious women, we intend to take our full share of responsibility for the honest administration of public affairs ... It has been said that the farmers have no great leaders. We are not looking for great leaders. Our wish is to make our organization absolutely democratic.[35]

For Griesbach, this was attainable only if the UFO remained a movement that granted women equal rights.

Griesbach had a great deal to say about relations between men and women, much of which challenged some of the farmers' basic assumptions on the subject. In an article entitled "Chivalry vs Equal Rights," she argued that many of those who called for equal rights and privileges were merely adopting a pose: "A very silly pose too, based on the groundless supposition that all women have or may have a lordly and chivalrous male who will protect her from all the rude world, (though if all males were lordly and chivalrous, where would the rude world be?) and who will provide her with a home in which she will be an honest-to-goodness 'Queen,' if only she will be satisfied with home for a Kingdom. (I would like to say 'queen-dom,' only there is no such word. Somewhat significant, eh?)"[36] For Griesbach, equal rights meant more than the granting to women of selected privileges by enlightened men.

Agnes Macphail frequently spoke on the plight of farm women and the indifference of their spouses. At a Ramsay UFO picnic in 1922, she observed that, although farmers worked hard, farmers' wives worked harder still. Macphail appealed to farmers not to deny their wives the kinds of labour-saving devices that they had for their own use. She proceeded to draw a dismal picture of conditions for rural women, commenting, "I don't think that marriage was meant as a burial place for women." She urged all farm women to support the UFWO club that had just been formed in Almonte. Much could be accomplished by the UFWO, and she emphasized that women should not wait for men to promote their cause. "Remember that the greatest dividend that will come is by doing something yourself."[37] Years later, speaking in Renfrew, Macphail asked an audience made up largely of women whether a week went by when they were not made to feel inferior to men in their own homes. For her, this was a clear case of domination, and it flew in the face of the rhetoric of cooperation in the farm family. In her speech, however, Macphail offered no solutions to these problems.[38]

At the local level, some UFWO members spoke in glowing terms of their husbands. "Ophelia" of Simcoe wrote, for example, that her husband "has always been my helper and pal where both flower and vegetable gardens are concerned."[39] Indeed, many farm men and women realized that mutual respect and assistance were required for a successful home; and genuine love and affection did exist in a number of farm families.

Despite examples of happy homes, UFWO members recognized that men were not always that accommodating or understanding when it came to women in the movement. Louise Collins had read that Agnes Macphail merely shrugged indifferently when men spoke of women as angels of the home. She, however, believed that men who characterized women in this way were gentlemen. "Where I live, if a woman writes a few lines on fair play for the farmers she is an Amazon."[40]

Some of the letters written to Griesbach on the topic of relations between men and women may have described personal experiences, and others may have put the problem in abstract terms. "Sister Ruth" of Lambton wrote, quite passionately, that when a farm woman went "to all the trouble of making over everything possible for her children's clothes that the farmer should appreciate her efforts to save enough that he would not spend thirty or forty cents every week on something [tobacco] which, as far as I can see, does him no good."[41] These issues, to be sure, were never resolved, but not for want of trying on the part of UFWO members.

Other Concerns

The sense of self-confidence engendered by being a UFWO member quite possibly enabled some women to address topics that were not usually discussed in public. Louise Collins had no qualms about admitting that she had once had a weakness for patent medicines that bordered on addiction, or confessing that she had at one time been swindled out of five dollars by a corn and bunion "specialist."[42] Writing to the *Sun* in response to a letter from a woman calling for the protection of girls, Alice Webster claimed that what girls needed were facts and not lies:

> Probably most of those who are mothers now can recall having gone through the process of having their innocence safeguarded with falsehood. I remember asking for facts and being told I was a "bad girl" for asking. But I kept on seeking. Biblical stories of gross immorality were eagerly devoured; other girls who bore the unsavoury reputation of "knowing too much" were eagerly questioned, and for the sake of the truth sought after I listened to foul stories and coarse jests that have stuck in my memory like a burn ever since, and all in the name of protecting modesty.[43]

Webster also noted that the author of an old medical text that she had consulted actually apologized for including venereal disease as one of his topics. What was needed, she claimed, was proper education. Perhaps children could be taught the facts of life by first exposing them to the process as it manifested itself in plant life. Next, animals might be used to describe reproduction. After that, human reproduction could be explained, and without embarrassment or resorting to falsehoods.[44] Responding to Webster, Griesbach called for more letters "on the important matter of sex education – real, genuine, rock-bottom, honest-to-goodness opinions." On another occasion, Griesbach said she thought it inconceivable that any mother could fail to inform their daughters of the facts of life. Yet she knew of some mothers who did not. The resultant ignorance, she argued, often led to disaster for young women.[45]

Writing from Simcoe in response to a letter that asked who knew more about mothering, a mother of six or a childless woman, K.N. Pepper argued that it ultimately depended on the training either may have had in child care. As a mother of eight, Pepper asked where a young mother, isolated from neighbours, was to get her knowledge in time of crisis: "I think it better to have the training before marriage, instead of experimenting upon the helpless children afterward. For me – I have one wee grave, which need not have been had I possessed the knowledge I should have had." Griesbach responded by asking whether there was "provision made anywhere in our whole social system for training for parenthood? Some people think it indelicate even to refer to the probability of future parenthood and the necessity for being prepared to discharge its duties."[46] Aside from the importance of the subject matter of this correspondence, it is unlikely that a letter such as the one Pepper wrote would have appeared in many urban dailies, or even in many rural weeklies.

Pepper's letter also demonstrates what can only be described as a disturbing trend on the part of women who corresponded with Griesbach – the tendency either to blame themselves for misfortunes beyond their control, such as in Pepper's case, or to denigrate themselves and their letters. "Cassie" of Simcoe concluded one of her letters by expressing the hope that Griesbach would not throw it in the trash, although she believed "it would feel more at home there than in the page with those other sisters' splendid letters." Numerous examples of these self-inflicted insults can be found.[47] Although some progress had been made in terms of women realizing their self-worth and their right to dignity, many of the old trappings remained. Clearly, even UFWO members continued to suffer from low self-esteem. Moreover, it may well have been the case that some were intimidated by articulate, self-confident women such as Griesbach.

THE FARM ECONOMY: WOMEN AND MARKETS

As I noted earlier, rural society was stratified as far as wealth was concerned. Some farmers were quite prosperous while many others barely managed to survive from year to year. The issue of rural poverty, or at least relative poverty, was a subject addressed by several women. Writing to Sister Diana, "Louise" of Stayner noted that she had recently been asked to complete the income tax return for the family, "as my time was not as valuable as that of the others." She set about the task with great seriousness, committed to sending the government every last dollar it was entitled to, because being a member of the UFWO meant being honest and giving all a "square deal." In the end, it turned out that the family did not owe the government a cent, because of the farm's low net income. "I think the figures will convince any right-thinking person that we are not on the farm for what there is in it but only because we like lots of room and fresh air." "A.A.G." of Simcoe claimed she knew of many farm families who were barely able to meet their expenses, much less have any money for luxuries. Yet these same people still "upheld their [political] party to their own hurt."[48]

Given the precariousness of the farm economy, it was important that farm women be involved in production. According to some scholars, however, the specific functions women assumed changed over time. In her study of farm women's labour, Marjorie Griffin Cohen argues that, as Ontario farmers turned away from wheat production and moved towards mixed farming, they became more involved in areas of production traditionally undertaken by women such as market gardening, fruit growing, poultry raising, and dairying. In Cohen's view, the growth of farm activity in these areas presented a paradox for Ontario farm women. Although women's participation in production actually increased in response to market forces, "there were also forces which ultimately would restrict that participation, at least in the agricultural sector itself." Focusing on the role of women in dairying, Cohen argues that the dairy industry became increasingly capital intensive, with the result that men – because patriarchal structures were firmly established and because governments tended to support "only male efforts in the industry as it became 'big business'" – effectively took over most operations.[49] Cohen's argument is a compelling one, and she has ably demonstrated how men appropriated the dairy industry once it became a significant part of the family income.

There are nevertheless some problems with Cohen's position. By focusing on dairying – a component of the farm economy that she admits was unique – she has paid insufficient attention to the areas of the farm economy that remained under the control of women. By ignoring these other areas, moreover, Cohen does not examine how women clung to various aspects of farm production. Evidence strongly suggests that women did so, and that the UFWO attempted to legitimize women's efforts in these areas by integrating them into the UFO's cooperative enterprises.

Women held on to many aspects of farm production that were aimed at outside markets. At a Perth UFO picnic in 1917, Mr Marcellus of the Live Stock Branch of the federal Department of Agriculture decried the fact that so many farmers in the area let their wives run the entire poultry operation of their farms.[50] "Sister Susan" of Simcoe informed Griesbach that in 1917 she had over sixty hens from which she sold close to four hundred dollars' worth of dressed poultry and eggs, in addition to what was consumed by her family.[51] Such activity on the part of farm women was often cooperative in nature, an approach that was encouraged by various authorities. In 1920 Meta Laws of the central UFWO was sent to Lanark to speak to women on the benefits of cooperative buying and selling.[52] Alice Webster of Creemore noted in August 1920 that many farm women were busy picking berries and fighting weeds in their gardens: "Yet when we take our berries to town we are told not to take them on Thursday because the [town] women are busy preparing in the morning for the weekly joy-ride ... and not to take them on Saturday because there is extra cleaning to do and they don't propose to do anything in the afternoon but dress up and go out."[53] Aside from the financial considerations, Webster pointed out that farm women would also like to take a few afternoons off, but under the present conditions it was impossible. At the inaugural meeting of the Almonte UFWO in 1922, local Agricultural Representative Fred Forsyth discussed the formation of an egg circle, a group in which members pool their eggs and sell them cooperatively.[54] Two years later, the UFO and UFWO in Lanark combined to organize a campaign designed to encourage cooperative marketing of produce.[55] One feature of the campaign was to encourage the formation of county egg circles as part of a provincewide egg pool. Mrs John Stewart of Appleton was given the task of explaining the plan to UFO and UFWO members. Stewart asked if farmers were foolish enough "to make our children stay on the farm when wheat, which is raised at a cost of $1.35 a bushel, sells for eighty-five cents?" In short, many women who continued to market their produce in fairly significant quantities sought to legitimize their efforts through cooperative activity.[56]

Much of the produce marketed by women was perishable and therefore vulnerable to the dictates of middlemen. Griesbach frequently argued that women often marketed their produce and saw little or no profit for their efforts: "I know that this is done right straight along. I do it myself, and I grind my teeth in helpless rage as I do it. It is produce which ... one cannot hold for higher prices, and the middleman and the consumer are perfectly aware of that."[57] Consequently, even though rural women continued to be active, they were in a more precarious situation than men, who could hold on to their more durable produce in an attempt to secure better prices.[58]

POLITICS AND THE 1919 UFO PROVINCIAL ELECTION VICTORY

Like their UFO counterparts, UFWO members initially stressed the need for political action, hoping to solve many of the problems they perceived by electing representatives from "the People." The enfranchisement of women enabled them to pursue political change enthusiastically through the UFO and Progressive parties. In addition, many farm women who had been politically inactive in the past worked to develop an understanding of politics and to put forward remedies for the ills they perceived, all in a remarkably short period of time.

After Beniah Bowman's Manitoulin by-election victory, farm women joined farm men in calling upon other constituencies to field agrarian candidates. As Simcoe's "A.A.G." wrote, "We should have our own farm representative and stick to him – no party politics. Then a man working for his own welfare will work for ours, and a farmer knows our needs, not lawyers, doctors, or inspectors running the country to fill their pockets ... The quicker we follow Manitoulin's lead, and work for ourselves ... the better."[59] If anything was to be done to benefit Ontario farmers, she maintained, it would be done only if farmers were elected to legislatures and if they played a role in decision making.

The 1919 UFO victory instilled in many farm women a feeling of self-confidence, and it strengthened their conviction that the movement would lead the province to societal betterment. Anna Elexey Duff of Lanark went so far as to write a poem and to illustrate it for the *Almonte Gazette*. The cartoon that she drew featured a farmer saying "Don't fear, I'm at the rear," as he pushed a wheel barrel containing a returned soldier and several full sacks of produce. The poem read as follows:

When we women drop the ballot, it will then be time to note
What the Country will answer, when we vote, yes when we vote
When we women drop the ballot, drop the ballot one by one
"Equal opportunities for all," "Special privileges for none"
When we women drop the ballot, it will then be time to stop

All intriguing Legislation from the bottom to the top
We will vote for honest labor in the Country and the Town
For characters of integrity make honor and renown
Loyal Legislator in the House on Parliament Hill
"Equal commercial values," will cure economic ill
For the things that are, and the things to be, in Canada home made
When we vote in Federal Government, and on the Board of Trade
On the Board of Education there will be a new trustee
For moderation in taxation, a National Policy
For Patriotism and Production always go hand in hand
With Temperance, and the Light of truth to cover all the land
Our vote is our protection, we will drop the ballot right, every one
For a "peaceful prosperous Canada," and a "Progressive Farmer's Son."[60]

In these lines one finds not only what were perceived to be the problems of society but also an optimistic sense of what would happen to these concerns now that the UFO was in power.

Louise Collins wrote to Sister Diana not long after the UFO victory in response to a letter in the urban press that blamed women for the Conservative defeat: "Oh! When will we poor long-suffering women cease to get the blame for every great calamity that falls on man? Mr Ray [the author of the letter] says that it was the women's votes that defeated the ... Government. Well we take the blame."[61] Collins also claimed that Ray had stated that women were so ignorant of political affairs they did not know the difference between a Grit and a Tory. She replied that women were, in fact, more intelligent because they realized that there was no difference. Later, Collins warned Ottawa to "take a hint ... and throw open the doors and windows at the Parliament Buildings and let in lots of fresh air, for there is going to be a real house cleaning." She also wrote that her local Progressive candidate, T.E. Ross, could rest assured that right would triumph over "that $1,000,000 government campaign fund."[62]

In subsequent correspondence Collins noted that many people objected to women participating in politics. She admitted that she too had objected until the UFO came along, and that she considered herself to be a progressive thinker because she had been able to change her mind. Buoyed by the election victory, Collins concluded one of her letters with a warning to potential candidates that they should not forget the promises that they had made during the campaign: "If they do, we women will jog their memories with their votes the very first chance we get."[63]

There were, as well, some UFWO members who were sceptical of farmers becoming involved in politics and of the role of farm women in it. "An Old Woman" in Simcoe felt that women could achieve more good by training their children to know right from wrong than they could by "trying to make the whole world right again by marking a ballot for a political party." She asked what the difference would be if the farmers actually formed their own party: "Will we avoid the mistakes of the old line parties; and will we let townspeople vote for our candidates? If we don't we're not logical, and if we do, we're not any better off than we were." Having made these observations, she admitted that she certainly would vote for a farmer candidate, because she did not wish to waste her vote by voting for the old-line parties, or by not voting at all.[64]

Interestingly, the correspondent apologized to any farm sisters she may have offended by writing this letter but then pointed out, probably referring to wartime censorship measures, that "there is no law now against expressing one's views." Although this woman may have exaggerated the government's repressive actions towards farmers during the war, there is evidence that state authorities watched farm groups with the express intention of catching them in committing seditious libel. Not long after it was formed the UFWO club in Milton, for instance, became the subject of an investigation by provincial officials. It was believed that seditious utterances were being made at club meetings, including suggestions that N.W. Rowell (president of the Privy Council and vice-chairman of the War Committee of cabinet) be taken out and shot, and assertions that farmers were not bound to follow any regulations set by the food controller. The attorney general's office arranged for a woman to attend a meeting of the group so that she could report on what had transpired. Evidently she was discovered as she attempted to enter the hall, and she was told that she was not welcome. Undeterred, the Attorney General's solicitor suggested to the local Crown attorney that the women who spoke in a seditious manner should be told "that they must keep entirely quiet or that stringent measures will be taken to quiet them."[65]

Occasionally, UFWO members actively participated in election campaigns, although many of the women speakers during the campaigns were drawn from the UFWO élite. During the 1919 provincial campaign, for instance, Mrs G.A. Brodie of Newmarket spoke at a political rally in support of East Simcoe UFO candidate J.B. Johnston. Even before Johnston had been nominated, Brodie had spoken at the UFO convention in Orillia. She pointed out that women never had "been paid for what they have done. We are penalized for raising children. The more we have the more we pay. Children are taxed before they come into the world. The cradle is taxed, and so are the bootees and other things they wear, while my lady's diamonds are free. Governments can conscript life, but it's like pulling teeth to make them conscript wealth."[66] During the campaign itself, however, Brodie made few references to women's issues. Instead, she addressed issues that were more relevant to men, or to men and women jointly, such as giving control of resources to the People, increasing farm income, and eliminating government policies that favoured the business class.[67]

A month after the UFO victory the UFO's Union Branch held a joint UFO/UFWO banquet in Collingwood. Brodie was one of the guest speakers who addressed the approximately 250 people in attendance. She informed her audience that, although the election presented new opportunities for farm men, there were also new opportunities for women "who were not willing to sit at the feet of the men and learn, but to stand shoulder to shoulder assuming a fair share of the responsibilities of the hour." Women's accomplishments during the temperance referendum campaign demonstrated their power in exerting influence in political matters. Brodie also informed those in attendance that the next federal election would be fought over the tariff, and that in preparation for the campaign, women should "study the tariff and see how you pay seventeen to forty-two and a half percent taxes on everything brought into the home."[68]

Even after the euphoria of the UFO's 1919 provincial election victory had died down, some UFWO members continued to be driven by the ideal of democracy and attempted to implement it elsewhere. Alice Webster attended a school trustee meeting and was surprised to hear for the first time "real democracy advocated from any platform other than the UFO." The speaker at the meeting argued that changing the schools was in the hands of the people. He went on to describe the kind of school he wished to see. For him, the ideal school would be one in which the "history taught would not be of kings and queens, but the struggles of the people for freedom."[69] Webster heartily agreed with this vision.

THE 1921 FEDERAL ELECTION

The 1919 provincial election represented the first time farm women actively took part in an election campaign, albeit on a limited scale. By the time a federal election was called in late 1921, however, many UFWO members were eagerly awaiting the opportunity to participate and work on behalf of Progressive candidates.

During the 1921 campaign working women from town and country united, as the men did, to fight those who in their view controlled

Canada's wealth. In North Simcoe Miss Mary McNabb, vice-president of the Ontario section of the Canadian Labor party, spoke on behalf of Thomas Swindle and called for a union of producers "of all kinds against those who only own wealth." The ideal that every woman in Canada should pursue, she believed, was to have a government led by producers. Referring to the Drury government, she said that it was a "humane government the people had placed in power ... If you were to die tonight your wife and children would be protected. Your wife would be paid a salary so that she might look after the bringing up of her children." McNabb also stressed the importance of the Progressive party's call for maternity benefits and free hospital treatment for expectant mothers.[70]

There were occasions when women drawn from the rank and file of local UFWO clubs were given the opportunity to speak publicly on election issues. In North Simcoe in 1921 a woman identified only as Miss Collins was given the task of introducing Agnes Macphail at the meeting to select the Progressive candidate for the riding. She warned her audience that something happened to men once they attained power, and it was the duty of "the People" to watch them closely so that the power was not abused. She also warned against sending "yes-men" to Ottawa, who would agree with their party's policies even if it meant ruin for their constituencies. According to Arthur Meighen, she claimed, the tariff was the main issue of the election. Collins disagreed: "The issue is, shall the people rule the people, or shall the people be ruled by the moneyed class of this country? There are many other issues in this campaign, and the efforts will be made to pull the wool over your eyes ... I heard one prominent man say that politics and patriotism were one and the same thing. Heaven deliver us."[71]

Space does not permit a discussion of the impact Agnes Macphail had on farm women in Ontario. Described by one Simcoe County UFWO member as "a real female Moses to lead us out of bondage," Macphail was an immensely popular figure who enjoyed considerable attention in both urban and rural newspapers. "Peggy Rambler" of Simcoe County wrote with pride after hearing Macphail speak in Stayner, "More power to your arm and to your tongue, dear Miss Macphail. A great many of us farm women and Sun Sisters are watching your career closely."[72] Although, as Rankin argues, Macphail often submerged gender issues and focused instead on matters of interest to both men and women, she referred occasionally to problems that were particular to women, and she remained a source of inspiration for women across Canada.

As with other parties, farmer politicians often made references to the role women played in the movement. Visiting Smiths Falls during the 1921 federal election campaign, T.A. Crerar pointed out that it was the Progressives in the West who had been the driving force behind securing the franchise for women. He went on to argue that "you cannot have a sound government unless it is conducted on a high moral plane and women will insist on a higher level ... and will not condone wrongdoing. The public man who uses his office to enrich himself or his friends is not going to have a very healthy time in the future with the women of Canada."[73] Significantly, Crerar made no reference to the desirability or even the possibility of women running for office themselves.

Crerar did, however, allude to the notion that women, now enfranchised, would elevate the dirty and corrupt world of politics with their spiritual purity. This theme was advanced by others as well. Speaking at a Lanark UFO convention, North Lanark UFO MPP W.I. Johnson congratulated local women for the interest they were taking in political affairs. In his opinion, women were going to be "one of the greatest factors in the cleansing of the political machine and in the cleansing of the policies that controlled this country." Clearly, the notion that women were morally superior to men was alive and well in Lanark.[74]

In Lanark, women played a greater role in election campaigns than in Lambton or Simcoe. Only in Lanark did the Progressives' county political association assemble a group of women directors, known as the Women's Progressive Club.[75] The directors of this club were determined to get local women interested in politics, even if they supported other parties. A room was set up in Smiths Falls for women during the campaign, and meetings were held there to explain the issues. The meetings were conducted as open discussions, and women affiliated with all parties were invited to attend. Women with differing political views were assured that their votes would not be solicited at these meetings. Meanwhile, Lanark women often came out in large numbers to hear the local Progressive candidate.[76] Even more to the point, women often took active roles in campaign meetings.

By the time the 1921 federal election was called the old-line parties had realized the importance of appealing to women voters.[77] The message presented by women in these parties, however, differed considerably from that of the United Farm Women. In Simcoe, for example, independent Liberal candidate Manley Chew employed the assistance of Mrs W.J. Stevens of the Toronto Women's Liberal Club, who told her audience that women should take an interest in the tariff issue because it took "such a large shot out of the pay envelope" of their husbands. A local woman, Mrs J.G. Needham, also spoke on Chew's behalf, focusing on the dishonesty and extravagance of the Meighen government. The only reference she made to women was to suggest that the Meighen government kept food prices artificially high with excessive tariffs.[78]

In Lanark during the 1921 campaign John A. Stewart, the Tory incumbent and minister of Railways and Canals, often had his wife address the women in his audiences. At a Pakenham meeting, Stewart's wife admitted that she was not a suffragette: "I did not want the vote. I do not think the women of this county were prepared for the vote but we have got it. It is now up to every woman to play her part." Regarding the tariff, she noted that women made up some ninety percent of the consumers of the country: "The men made the money and the women spent it, but they had been spending too much of it across the line." If women would only take the time to ensure that the goods they purchased were Canadian-made, most of the problems associated with the tariff would be solved. Later, Stewart spoke in Perth with Mrs Agnes Munro of Winnipeg. Munro argued that it was difficult to assimilate immigrants, "yet these foreign Canadian citizens had a vote ... so it was up to us to educate them along proper lines of citizenship."[79]

Stewart's and Munroe's comments were too much for at least one woman to take. Mrs J. Stewart of Appleton had never been to a political rally until mid-November 1921, the one in which Mrs John Stewart and Munroe spoke. "Of all the rubbish, of all the flag-waving, belittling the loyalty of Canadian citizens and throwing mud at the United States," said Stewart at a Progressive meeting. She addressed the cry of "drive them out" when both women referred to immigrant women in the West. "Had anyone ever heard such rot? They were to be driven out because they didn't appreciate ... Meighen." Stewart went on to note that women made up sixty-five percent of the electorate, and that it was no wonder that politicians were trembling at the prospect of what women might do with their votes. She then speculated that, if the women had had the vote ten or fifteen years before, they might not have exercised it; but "the war was a great eye-opener and women had begun to do their own thinking. They had come out in Ontario when a great moral issue was at stake," and they had been active in the West as well. Stewart claimed that Mrs John A. Stewart had gone to Carleton Place and Almonte and assembled loyal women and requested them to ask every man and woman how they intended to vote. "Did you ever know of anything worse?" Stewart maintained that that kind of act was at the very foundation of dirty politics, a brand of politics that was going to cease once the Progressives were in power. She appealed to her audience not to dishonour themselves: "Explain matters to them, but don't ask them how they are going to vote." She had no sympathy for those who claimed that dirty politics were merely "part of the game." In concluding, she remembered how John A. Stewart had said in Carleton Place that agriculture was the basic industry in Canada but had "then gone on to extol the manufacturers and sneer at farmers."[80] All in all, a highly impressive showing from a person who had not been politically active a few weeks prior to this meeting.

R.M. Anderson, the Progressive candidate for the riding, spoke after Mrs Stewart at the meeting and devoted a considerable portion of his speech to women. He believed that women were approaching this campaign in a different frame of mind. They were not so biased, and "all the Progressive party asked was that the ladies in studying the questions of the day would be guided by their own intelligence; that they would study the questions for themselves and come to their own conclusions." Anderson resented the way the other parties brought in women from outside the county to show the women of Lanark how to vote. On another occasion, he claimed that women deserved much of the credit for breaking down the old party loyalties. He believed that men were prone to support the parties their fathers did: "Would the ladies do that? No. Their very curiosity would lead them to ask questions and in that way facts would be brought out and we would get better and better government."[81] Anderson's use of stock phrases aside, it was clear that in Lanark, women were not ignored in 1921 to the extent that they had been in 1919.

Stewart was not the only woman to express disgust at the way in which racial and ethnic questions were addressed in the 1921 election. Alice Webster noted that federal Tory candidate Col. J.A. Currie was filling Simcoe with warnings that Quebec might end up ruling the country if the Tories were not given a majority. Currie also alluded to the great danger of Quebecers and farmers forming a coalition government. Webster later observed that, in the election, there were "those who – accusing everybody outside the Tory party as being unpatriotic – are making supreme efforts to disrupt Canada with race and religious hatreds." As such, farm women had to "try to stem the evil tide and promote a united Canada."[82]

The Influence of the State

There were times when the state attempted to influence the UFWO. In Carleton Place, for example, a social meeting of the town's UFO and UFWO featured a Miss Williams of the Kemptville Agricultural College, who offered her services "to organize tennis clubs, basketball and to conduct classes in Domestic Science." Also featured was Professor Bell of the same institution, who spoke of the splendid work of its teachers and of the "value to the boys and girls on the farms of a course at this College."[83]

In March 1922 a UFWO club in Lanark was established at Almonte. The organizational meeting was addressed by Mrs George Buchanan of Appleton, John T. Somerville of Middleville, and Fred Forsythe, the local agricultural representative. Somerville and Forsythe spoke on the question whether an egg circle should be formed with Almonte as its headquarters.[84] The Union Club 604A in Simcoe was visited regularly by Alan Hutchinson, the local agricultural representative, who in one such meeting showed moving pictures and gave a talk on canning. Hutchinson's successors continued speaking at UFWO meetings, often on domestic science topics.[85] The Pakenham UFWO listened to a talk by the home-nursing instructor for the agricultural short courses that were being held in that town.[86] At the central level, a Mr Benson of the Ontario Department of Agriculture addressed UFWO members at their 1923 annual convention and instructed them on the packing and candling of eggs – whereby the eggs were held up to a candle to test for freshness – and on the formation of egg circles. At the next annual convention, federal officials were on hand to demonstrate the advantages of forming these circles.[87]

Relationship with the Women's Institute

No study of the UFWO would be complete without a discussion of its relationship with the Women's Institute, the most blatant example of state involvement in women's organizations. Sponsored by the provincial government, the WI was designed to encourage continuing education in areas such as domestic science and health, and to provide social and cultural programs for rural communities.[88] It was a highly popular organization and, as time went on, remained much more vibrant and viable than the UFWO. There was a tension between the two groups,[89] at least at the central level: at the local level, the UFWO was as often as not on friendly terms with the local Women's Institute.

Alice Webster was among those who were sceptical about the ability of the Women's Institute to further the cause of women. She had worked hard for the institute when it first appeared in her locality. In her view, its restrictions on the political and religious topics that might be discussed did not initially cause any great problems since "we were dealing with a class of women who were frightened when asked to express their ideas at a meeting." Gradually, however, these women began to ask increasingly difficult questions, and they gained the courage to speak more freely on many topics: "I have watched the growth and rejoiced. But now we have reached the point where barriers prevent further progress. Barriers erected by the Department of Agriculture, but more rigidly upheld by those who feel it their duty to shield a corrupt Government from criticism. Since women are now becoming citizens instead of wards, we must have a means whereby we may qualify for citizenship."[90] Since the Women's Institute was not capable of providing such means, Webster argued that it should make way for other groups, such as the UFWO and the Ontario Women's Citizens Association. As Louise Collins put it, "All honour to the Women's Institute, but you might just as well try to mix water with oil as to join the Institute and the UFWO. We might agree on the weather and the fashions; but what town women want to pay for butter and eggs, never." Clearly, Collins saw the institute as urban dominated with an urban-oriented agenda. A woman from Simcoe County, Mrs James N. Foote, proposed a motion at the first UFWO convention stating that there was a tendency on the part of the Women's Institute to "stifle any sign of independent thought on the part of women." Another woman agreed, stating that there was "no reason to hope for any improvement from a Government controlled institution."[91]

In many rural areas, however, most women clung loyally to the institute. In Simcoe County, for example, many institute clubs persisted even after the provincially sponsored Farmers' Clubs became affiliated with the UFO. Moreover, the local UFO/UFWO clubs and the Women's Institute occasionally worked together.[92] In 1921, for instance, a community hall was constructed in Eady, North Simcoe, as a joint effort of the UFO and the local Women's Institute.[93] At the annual East Simcoe UFO picnic in 1922, the UFWO and the institute held a joint tea, featuring such speakers as Mrs E.C. Drury.[94] In other areas, such as Harvie Settlement in Simcoe, the local Farmers' Club and the Women's Institute held meetings in concert well into the 1920s and managed to carry on good relations with local UFO and UFWO clubs.[95] In many cases, in fact, local UFWO meetings began to resemble those staged by the Women's Institute. The Wanstead UFWO spent a great portion of its monthly meeting in January 1925 quilting. At one of its meetings in 1926 the Forest UFWO discussed household duties for the month of March.[98] The Almonte UFWO maintained the "Janey Canuck" ward in the local hospital. The Guthrie UFWO in Simcoe County had each member donate a jar of fruit or pickles to a Barrie hospital in 1925.[97] Later, it decided to purchase individual drinking cups for students at two local schools and also to provide pictures for these institutions.[98] The suggested program for local UFWO clubs devised in 1925 by Mrs J.S. Amos, UFWO president, supports the contention that the UFWO gradually moved into areas traditionally operated by the Women's Institute (see Appendix T). Topics such as the ones outlined above were discussed at UFWO meetings from its inception. The point, however, is that they were addressed much more frequently as time went by.

By the mid-1920s the UFWO membership began to decline along with its effectiveness in the struggle for political, economic, and social equality. Although the softening of the UFWO agenda can be seen as a retreat on the part of the organization, it could well have been the case that the frustrations women experienced in political, economic, and social areas pushed them back to topics and areas of concern where they felt they had at least some agency. Many farm women returned to the Women's Institute. Indeed, as Rankin argues, they most probably did so because the institute was "an organization dedicated to legitimizing rural-women's domestic pursuits [and it was] already engaged in reformist initiatives."[99]

The reason why the UFWO was created in the first place, however, was not always forgotten. At a meeting of the Forest UFWO in 1925, Mrs Darville, district organizer, urged its members "to be interested in the problems outside the home such as the community, the school, and the province. To accomplish things ... we need to read and think for ourselves."[100] In addition, the chief speaker for the Carleton Place UFO picnic in 1925 was UFWO president Mrs J.S. Amos. Her message that day was reminiscent of the sentiments expressed in the early days of the UFWO.[101] By that time, however, most women were doing their thinking in the Women's Institute or on their own.

THE 1923 PROVINCIAL ELECTION AND POST-ELECTION POLITICAL ACTIVITY

By the time the 1923 election was called, the UFWO was not nearly as strong in its support of farmer candidates as it had previously been. At least this is what evidence from the *Sun* indicates. Only a few letters to the paper from women during that time made any reference to the election.[102]

There were, however, some exceptions. "Ophelia" provided *Sun* readers with her observations on the election. First, she scoffed at the man she overheard at the poll who could not understand why women did not vote for the same candidates as their husbands did. She also noticed that many people whose friends had benefited from the

Widowed Mothers' Allowance voted against the government that had legislated it (i.e., the UFO). Another person told Ophelia that the UFO government had not done much for farmers: "My reply was that they tried to benefit the people as a whole ... Who gave us improved educational means and school grants that enabled the farmer to pay teachers better salaries? Who gave widowed mothers' allowances, enabling them to keep their children in their homes ... Who gave support to the mother of the illegitimate children? Has not the guiltless child an equal right with other children?"[103] She also noted with disgust that in her community Drury was burned in effigy after the votes were counted. For Ophelia, women in the province now had to ensure that the gains they had made under the UFO were not eroded.

UFWO interest in politics did not, of course, end as a result of the 1923 defeat of the UFO. At the 1925 annual UFO convention, for example, Mrs C. Darville of Lambton moved to restore "the nomination of UFO candidates and participation in politics by the association." In 1927 Emma Griesbach asked whether farmers were going to continue to be "bamboozled by political hocus-pocus and claptrap" during election campaigns.[104] By that time, however, the farmers were of little significance in provincial politics.

Buoyed by the confidence they had gained while participating in the provincial and federal elections of the early 1920s, UFWO members remained active in the internal politics of the UFO. In Lanark they even stood for and held elected positions, unlike Simcoe or Lambton. In late 1922 local MPP Hiram McCreary nominated Edna Gardner, a schoolteacher in Ramsay Township, to the position of secretary-treasurer of the County organization. No other candidate was named. Gardner was reported to have been surprised to be nominated, but she accepted the position on the executive nevertheless.[105]

The election of women to the local UFO executive remained commonplace in Lanark for quite some time. In 1926 and 1927, for example, Miss Mary Lyle served as the county organization's secretary; in 1928 Mrs T. Armstrong held the position. In 1931 Miss Hazel Thom was elected secretary-treasurer of the county organization. This pattern was not in evidence in Simcoe or Lambton,[106] however, and the presence of women on the Lanark UFO executive invites some speculation. Perhaps women were more important to the farm unit in Lanark where, in many cases, everyone had to contribute to the family enterprise merely to get by. In addition, as noted elsewhere, Lanark displayed a greater radicalism than Lambton or Simcoe at times during the peak years of the UFO, and this radicalism may be reflected in the presence of women on the Lanark UFO executives.

THE DECLINE OF THE UFWO

What accounts for the decline of the UFWO? There were many factors, including the general decline of the UFO throughout the province. Another reason was the disillusionment experienced by women when it became evident that the UFO was not going to alter its patriarchal structure to any great extent. This was perhaps best summed up in Alice Webster's letter to W.C. Good protesting Emma Griesbach's removal from the *Sun* after the male-dominated executive of the paper (and of the UFO) had become uncomfortable with Griesbach's feminism. After describing all of the good that Griesbach had done for Ontario farm women, Webster wrote, "in the midst of our efforts to raise the standard of rural life you men have stepped out and struck a blow that sends us reeling backward. *Why?* Do you think the women of this country can put any faith in the sincerity of the UFO when you talk of giving equal rights to women, after you have cut us off from our leader?"[107] Despite the rhetoric of "equal rights for all," many in the UFO only paid lip service to the idea, and reacted harshly when any threat to the established order presented itself.[108] After witnessing this sort of behaviour, it is reasonable to assume that some UFWO members became disillusioned and left the movement.

CONCLUSION

It is difficult to assess the impact such organizations as the UFWO had on women in general. There is little agreement among those who study women in agrarian movements as to the effectiveness of these movements in advancing the quest for gender equality. From the preceding, however, a few general comments can be made.

Farm women – like their urban counterparts – experienced (and were cognizant of) the contradictions and confusions of early-twentieth-century Ontario. However, since they had recently been enfranchised, many of them looked towards a "new day" wherein they could play a meaningful role in the governing of the province. In concert with this notion, it appeared that new societal attitudes paved the way for farm women, as well as for women in general, to emerge as true equals in every respect to men. Expectations were therefore high.

In the case of farm women, there was already a tacit acceptance of their importance with respect to the farm economy. All members of a farm family, after all, worked so that the farm would be productive. But acknowledging the importance of women on the farm did not always mean that they were treated as true equals, and in most cases they were not. As a result, many of the women who joined the UFWO expecting to end their isolation, reduce their extremely heavy workload, correct urban misconceptions about rural life, and achieve equality were sadly disappointed. In many respects, then, the movement failed to fulfil its expectations.

Yet the situation was not as grim as this "failed outcome" suggests. The UFWO gave many Ontario farm women their first taste of what it was like to be a member of an organization and provided the vehicle through which they let others know how they felt about things that affected their lives. Moreover, the UFWO provided women who had never been allowed to participate actively in political affairs with the opportunity to learn about how the province and country were governed, and the chance to reflect on how conditions could be improved. Many women grew in self-esteem and self-confidence as a result of UFWO membership and were thus able to participate in the articulation of an alternative vision for society. For many UFWO members, the lessons obtained from their experience in the movement served them well in later life.

6 "To Eliminate the Capitalist and Profiteer Is Simple": Agricultural Cooperatives in Lambton, Simcoe, and Lanark Counties

> We do not even know who sets our prices, or why they do it or how. Only we're beginning to learn that these people whom we have allowed to do our thinking for us, have not been thinking for us at all but for themselves ... we have at this late hour discovered that we are not merely individuals with interests limited by the four walls of our homes ... and with this awakening has come, of course, the get-together idea, not *I*, but *WE*, and we are organizing.
>
> Margery Mills[1]

People form cooperatives for a variety of reasons. They may wish to fill gaps where capital is either weak or unavailable; to propagate a communitarian ethos where wealth is distributed according to effort and not capital; to meet the needs of members rather than the needs of the market; or to express antipathy to the excesses of capitalism.[2] At the root of all these motives is the desire to meet human needs on egalitarian principles.

In most cooperatives in Canada the way to achieve this goal in the late nineteenth and early twentieth centuries was to establish the cooperative on the Rochdale plan (that is, one member, one vote). Obviously, then, cooperatives were (and still are) formed in opposition to prevailing economic and political norms; "The principle of one member, one vote, of economic returns to patronage rather than capital, of open membership and of cooperative education, formed a conceptual whole which makes little sense except when seen as an alternative to economic and political institutions that sustained an undemocratic social order."[3]

Cooperatives are often seen as the cornerstone upon which agrarian populist movements were built. Lawrence Goodwyn argued that farmers' experiences with co-operatives "radically altered their political consciousness." Co-ops offered farmers a valid alternative to the capitalist economy in which they found themselves and were used as the "central educational tool" of the Farmers Alliance.[4] Historians examining agrarian cooperatives in Canada have arrived at similar conclusions.[5]

Of course, conceiving of an alternative to prevailing norms and then implementing such a vision would inevitably be met by subtle as well as overt opposition. Perhaps the best way to demonstrate subtle opposition – the power of hegemony – and its consequences for a mass movement such as the UFO is to provide a case study of its cooperative activities. Cooperation was a vital component of the UFO. And, as with other late-nineteenth- and early-twentieth-century populist organizations, the failure of its primary cooperative endeavour, especially when combined with political failure, had a devastating effect on the movement's ability to eliminate the inequitable power relations it perceived and advance its alternative political, economic, and social vision.

In some respects, co-ops affiliated with the UFO's companion organization, the United Farmers Cooperative Company (UFCC) did not fail. As will be seen, many local cooperatives met with a measure of financial success, and some of these profit-making co-ops outlived the UFO. Yet increased returns for farmers were only one of the many reasons for establishing UFO co-ops. By the mid-1920s, however, with the exception of some local cooperators, few UFCC organizers discussed anything other than the profitability of their ventures. Thus, UFO cooperatives failed, at least as alternative enterprises.

Co-ops of one form or another had existed in Ontario well before the emergence of the United Farmers' movement,[6] but the establishment of the UFO and the UFCC in 1914 further entrenched cooperative activity in the province. Consumer co-ops, producer co-ops, cooperative marketing associations, cooperative stores, and cooperative mills flourished in Lambton, Simcoe, and Lanark counties and indeed throughout Ontario. These enterprises allowed UFO members to experiment with alternative forms of commercial as well as social and political relations that were markedly different from the *status quo*. Whether or not one sees cooperatives as radical bodies is a matter of opinion. In the case of the UFCC, it is ludicrous to suggest that most members were bent on overturning all societal institutions. What I argue here is that cooperatives, by promoting community development, direct and meaningful democracy, and mutuality rather than the possessive individualism of capitalism, are *potentially* radical. In short, experience in a co-op may lead participants to question some of the fundamental "truths" of existing social, political, and economic relations. Yet by the late 1920s, agrarian co-ops, especially large-scale ones, were shadows of their former selves.

This chapter examines the transformation of UFCC and UFCC-affiliated co-ops from vehicles of potential change into something resembling capitalist enterprises. Four main arguments will be advanced: first, that the state played a significant role not only in encouraging certain types of co-ops (specifically, marketing co-ops) but also in shaping them to suit ends that were quite different from those of farmers; second, that farmers were adversely affected by the heavy-handed policies and increasing centralization of the UFCC; third, that the UFCC's leadership came to view cooperatives less as transformative agencies than as money-making ventures, and that they transmitted this message to rank-and-file members at every opportunity; and fourth, that despite these pressures, spontaneous local co-op activity persisted in all three counties during the 1914–30 period, as participants attempted to adhere to the professed ideals of the UFO.

INVOLVEMENT AND INFLUENCE OF THE STATE

There are few studies of the role of the state in the cooperative movement in Canada. Those that do chronicle the Canadian experience tend to focus on the grain-growers' cooperatives on the Prairies.[7] Yet the state's role, both federal and provincial, cannot be ignored when examining Ontario agriculture during and after the First World War. The federal government actively fostered the development of co-ops even before the war,[8] and by the mid-1920s federal authorities were involved in several initiatives. To cite but a few examples, the Dominion Seed Branch subsidized the Canadian Seed Growers' Association by some ten thousand dollars annually and provided up to two hundred dollars (with an additional hundred dollars advanced by the provinces) to assist with the costs of staging local seed competitions. In addition, for two dollars the branch would send a farmers' club or a local cooperative plans for the construction of a small seed elevator. The federal government was also instrumental in establishing the Canadian Cooperative Wool Growers in 1918 and for some time after that supplied expert graders to the company.[9] Moreover, "by legislation, regulation, supervision and instruction in the federal field during the past two or three years a distinct advance has been marked in connection with the grading and shipping of fruits, potatoes, eggs, poultry and dairy produce."[10] Much of this activity involved establishing and supporting producer cooperatives.

Some of the scant documentation regarding the federal government's role in encouraging producer co-ops reveals the motivation behind such support. Writing to the deputy minister of Agriculture in 1920, the Dominion Live Stock commissioner noted that Canada had developed some first-rate marketing organizations that had secured good returns for producers. The great problem, however, was to obtain the highest-quality product and to ensure high-volume shipments. To realize these goals the government must "concentrate upon the output of the individual community." The proposal he advanced was designed to "assist individual communities in the marketing of their stock and produce, to aid the District Representatives in promoting cooperative action for this purpose ... to improve the quality and increase the volume of a community through this means." Roughly one year later the Live Stock commissioner informed the deputy minister that the present system of marketing hogs in Canada was "one of the worst in the world ... packers have adopted a system of shipping with so little regard to uniformity respecting quality and weight."[11] Again the solution was to be found in consulting with such groups as producers' associations and with the Canadian Council of Agriculture so that a cooperative system could be devised.

Simultaneous with these efforts of the federal government, provincial officials were also active in assisting farmers throughout Ontario. Disseminating information was a major concern of the provincial government, and W. Bert Roadhouse, deputy minister of Agriculture, boasted that in 1917 nearly one hundred thousand pieces of literature were distributed to Ontario farmers.[12] In addition, the province provided speakers upon request to address a wide range of agricultural topics, and it offered short courses and frequently held demonstrations in many localities.[13] In Lambton (and elsewhere), local farmers were successful in securing grants of up to $350 from the province for the county's annual Corn Show. In addition, the province continued to encourage farmers to drain their land, and loans were established for this purpose and even for the acquisition of farms.[14]

The province also responded to problems that the Great War created for farmers. With global demand for wheat at an all-time high in 1918, the Ontario government purchased fifty thousand bushels of No. One Marquis Spring Wheat seed from federal authorities. The seed was sold in two-bushel sacks at an attractive price, and farmers' clubs could order carload lots. In carrying out this plan, the government achieved three main goals: the province had surplus wheat available for export at harvest time; by and large it was a standard type of wheat; and farmers believed that the state was working in their interests. Some farmers, however, protested that this scheme deliberately undercut the UFCC's seed prices.[15]

A severe winter in 1916–17 led to a shortage of seed of all types. In response to the crisis the Ontario Department of Agriculture secured a supply of several types of seed, which was distributed to farmers at cost. When it was discovered that many farmers were unable to pay up front for the seeds, the government arranged with the Canadian Bankers Association and the Organization of Resources Committee to advance up to two hundred dollars for this purpose. Some farmers were critical of the program. In a letter to the *Sun* John M. Houldershaw of Simcoe County wrote, "I do not know of any farmers that require assistance ... to buy seed ... their credit is usually good, but they do not want to buy seed potatoes when it requires thirty dollars to plant an acre. Sometimes the crop will not sell for thirty dollars."[16]

In addition, during the war the province launched campaigns to encourage urban residents to assist farmers at harvest time.[17] Most of this work was accomplished under the Patriotism and Production campaign, a federal-provincial effort designed to keep agricultural and production levels high.[18]

More importantly, authorities in Ontario were very active in their support of marketing cooperatives. The provincial Department of Agriculture set up the Cooperation and Markets Branch in 1914, and legislation was passed to facilitate cooperative marketing, including legislation that provided loans to co-ops for the purpose of establishing seed cleaning plants and potato warehouses.[19]

One blatant example of the province's interest in cooperative marketing appeared in a government advertisement in a 1917 issue of the *Sun*. The advertisement used one of the more popular methods of getting messages across to readers – the political cartoon. Under the headline "Organized Marketing on a Business Basis Means Increased Profits for You," the cartoon featured a farmer carrying a bag representing the season's crops. There was a hole in the bag, and the produce that was falling out was being eaten by four fowl, named "superfluous middlemen," "poor storage," "bad packing," and "individual selling." After presenting the standard information regarding the benefits of cooperative marketing, such as improved quality control, better packing, and so on, the advertisement concluded with "Cooperation, in short, PAYS AND PAYS WELL."[20]

Provincial Department of Agriculture officials recognized the benefits of using the local press to deliver their message of cooperation, and it was used frequently. Week after week farmers were exposed to columns supplied to local newspapers by the department. A typical column can be found in a Smiths Falls newspaper,[21] and some insights can be obtained from an examination of what was said and, equally important, what was not said in the article.

Entitled "Cooperative Selling," the column began by observing that agricultural products made up a great percentage of Ontario's exports. Hence, "we have to see that our agricultural products going to the markets of the world go ... in the shape and form demanded by the markets we are attempting to gain." Grading was necessary, and produce had to be of uniform quality to be acceptable. Cooperative marketing would help ensure that these standards were met. It would also stop the dumping of produce on the market at harvest time, which lowered the price and resulted in enormous waste. According to the article, only the speculator benefited from dumping, and farmers, not those who speculated on the market, were to blame for the situation: "The speculator does not break the price, the farmers do this themselves by dumping their product one against the other, making it possible for the speculator to watch the fight ... and then step in and take the spoils at his price." In other words, farmers, and not the system that was conducive to speculation and dumping, were at fault.[22]

Further on in the article it was noted that a co-op was more likely to succeed if it focused on a single commodity, such as fruit, grain, dairy products, or livestock. Particular difficulties had to be overcome for each commodity, and to mix activities spelled trouble for farmers: "Organizations that have attempted to handle the marketing of numerous lines of farm crops have generally been unsatisfactory in that the divided interest of the cooperative is destructive to success." Perhaps the author had the UFCC in mind when writing this passage.

The article was silent on other benefits of cooperation. That co-op marketing allowed for some measure of local control, or that it helped to foster a sense of community and self-help in local farmers, went unmentioned. Also ignored were the democratic tendencies of cooperation and the perceived long-term effects of large-scale cooperative activity, such as the gradual elimination of middlemen and cut-throat competition. Yet these were the themes that characterized the rhetoric of the idealistic UFO rank and file.

State support of cooperatives was deemed important enough to be mentioned in election campaigns. James S. Gould, the Conservative candidate for South Lanark in the 1919 Ontario election, devoted considerable space in his advertisements to describing the Hearst government's commitment to encouraging cooperation. Enabling legislation had been passed, the Cooperation and Markets Branch had been created, and assistance had been granted to a number of producer groups. In addition, for some time the government had carried out a campaign "showing the value of cooperation among the farmers and has assisted in organizing upwards of four hundred Farmers' Clubs in the Province, giving full information as to the organization and business, *but leaving the conduct of the business affairs to the farmers themselves.*"[23] Gould then boasted that "it is not too much to say that seventy-five percent of the cooperative effort of the farmers ... is due to the foundation educational work which has for years been carried out by the Department of Agriculture." And in case anyone got the wrong idea, Gould reminded his readers that "students of the subject know that farmers' cooperative organizations in other countries are kept free from politics."[24]

After the UFO's victory in 1919, the provincial government stepped up its efforts to encourage co-op marketing among Ontario farmers. In 1922 Agriculture minister Manning Doherty arranged to have Aaron Sapiro, a co-op marketing expert from California, travel throughout the province to promote cooperation in rural communities.[25] In one speech given near Lanark, Sapiro extolled the advantages of cooperatively marketing cheese. After pointing out the benefits that accrued from this method of selling (standardization of product, effective grading, better distribution, and so on), Doherty took the stage and announced that the province had no intention of controlling or even directing cooperative selling, insisting that co-ops had to organize and run their businesses through their own membership. The government wished only to assist "in organizing and provid[ing] such legislation as may be necessary." Doherty made it clear, however, that the aim of the province was "to organize all farm industries in joint-stock companies operating strictly as sales and manufacturing pools," which contradicted his earlier statements about farmers running their own affairs and setting up co-ops based on local conditions and needs.[26]

The state's interest in cooperatives – and the character of that interest – can also be seen in the report of the provincial Agricultural Enquiry Committee, which was struck in 1924 to "study all matters concerning the social, educational and economic conditions surrounding the agricultural ... industries of the Province." One of the main concerns of the committee was the marketing of agricultural products for domestic and overseas markets. Regarding livestock, the committee members conceded that the situation was not dire; however, "in the marketing of cattle ... cooperation will have to be depended upon for future headway." The same prognosis was made for field crops.[27]

The committee went further than most other government bodies in that it also supported consumer cooperatives. After several witnesses complained about the high price of agricultural implements, for the most part blaming implement agents and the high commission they charged, the committee members concluded "that the practical remedy ... is the formation by the farmers of a cooperative buying agency." Unlike their recommendation concerning producer co-ops, however, they did not propose that the government get involved in such a scheme.[28]

By the time the Agricultural Enquiry Committee hearings were underway, at least a few prominent UFO members had begun to express their concern about the activities of the state with respect to co-operatives. In its statement to the committee, the UFCC argued that co-ops often worked best when they emerged from the people themselves. Granting this, it was the company's opinion that the government "should in no way attempt to force, control or direct the trend of the cooperative movement ... [state activity] should be limited to providing facilities for securing the fullest and most reliable information regarding cooperative practice." The UFCC noted that the provincial government had recently provided a group of turnip growers with a subsidy of one thousand dollars so that representatives from the group could investigate turnip markets in the United States. If the UFCC was to do the same, it would have to be completely funded by the company.[29]

Why, one might logically ask, did the state take such an interest in the activities of marketing co-ops in Ontario? In a capitalist society the state, quite naturally, tends to serve the interests of capital. Although there is considerable debate as to how the state accomplishes this, it is generally agreed that it will try to avoid actions that work contrary to the long-term interests of the capitalist class. It is here, as one might expect, that cooperatives – which generally operate on principles that run counter to the capitalist model – present problems. Yet certain types of co-ops, namely, producer co-ops, were highly beneficial to the state. Most states depend to a high degree on trade (and Ontario was certainly no exception), and marketing co-ops presented policy makers with an ideal opportunity to further the goal of increasing both domestic and international commerce. Clearly, for the federal and provincial governments, the reason for encouraging producer co-ops was based in no small part on the fact that they facilitated and improved international trade.[30] In a country whose agricultural exports amounted to a significant share of its total exports, this was an important contribution. By relying upon marketing co-ops, a standardized grading system could be implemented –

thus ensuring high quality – and produce could be assembled in bulk quantities – thus facilitating large international shipments.[31] It is not surprising, then, to find that the state, from very early on, encouraged farmers of the province to participate in producer co-ops. It was a fringe benefit for the state that the public believed that it was serving the interests of "the People."[32]

One problem, however, was that producer co-ops were not the only type of cooperative activity that proved attractive to farmers. Consumer co-ops also had great appeal; but since they ran counter to the goals of a capitalist economy, state authorities were somewhat less than enthused about their existence. They could not take direct action and eliminate them through legislation, but they could be passively obstructionist. It certainly proved a struggle to obtain legislation for consumer co-ops that was even remotely favourable, and neither federal nor provincial officials went to any great lengths (as they had for producer co-ops) to support or guarantee the survival of consumer co-ops in Ontario. By neglecting, even disparaging, consumer co-ops the state was clearly acting in its own best interests, interests that were at odds with the actual needs of farmers.[33]

Two final points regarding state support of marketing co-ops should be made. First, contrary to the common belief that the state involves itself solely to benefit farmers, V.C. Fowke argues that the main beneficiaries of supportive state policies are consumers, who enjoy lower food prices. Second, federal and provincial officials often claimed that ensuring high quality and standardized produce would mean increased profits for farmers. This is a debatable assertion: if grade A hogs sold at three dollars more per hundredweight than grade B hogs, and five dollars more than grade C, and if one million of each were produced in a given year, then it would seem logical to assume that, if farmers produced nothing but grade A hogs the following year at the same rate of production, an additional seven million dollars would be realized. As Fowke argues, however, this is not the way the price system works. In reality, if all hog farmers produced grade A hogs, then the market would be flooded with these hogs and the price would be forced down towards grade B or even grade C levels.[34] It is not known whether or not state officials knew this, but even if the state policies had worked perfectly, farmers would not have realized significantly higher returns for their produce.

LOCAL COOPERATIVES – ESTABLISHMENT AND INITIAL OPERATIONS

As seen, local and even UFCC-directed co-ops were never entirely free from state influence. This, however, did not preclude spontaneous and creative action on the part of UFO members. Moreover, although the UFO victory in 1919 resulted in a rapid expansion of the co-op movement, expansion merely augmented the considerable activity that was already evident in all three counties under consideration. An examination of local co-ops in Lambton, Simcoe, and Lanark provides insight into how such groups, over time and under pressure from the state and the UFCC, began focusing more on profitability than on the less-tangible aspects of cooperation.

In Lambton, local commodity-specific associations already existed, among others, along with a farmers' mutual insurance company. There was also a countywide organization known as the Lambton Farmers' Cooperative Society. Formally established in 1915 with the amalgamation of a number of farmers' clubs in the county, the society had several directors who later became prominent local figures in the UFO.[35]

Early on, members were warned that they had to exercise caution when discussing UFO business transactions publicly. Addressing farmers in Lambton in 1915, Anson Groh, then with the UFCC, offered to take orders for binder twine and advised his audience that it would be unwise to make the price known "as the dealers would then undersell them in an effort to break up the [society]." Some seventy-five tons of twine were ordered by the two hundred or so farmers in attendance.[36]

It was also during the formative years of the UFO that Lambton farmers received a taste of the attitude of the provincial government. Speaking at a meeting of farmers at the Lambton Corn Show in 1915, F.C. Hart of the Cooperation and Markets Branch gave the farmers "a few plain truths" about cooperation. He noted that co-ops – particularly consumer co-ops – required adequate capital if they were going to succeed; reminding the farmers that the middleman's margin was not all profit, he stressed that there were overhead costs that had to be taken into account. As if to solidify his role as wet blanket, Hart then remarked that the farmer was often "not a good businessman."[37] Later that year, Lambton Agricultural Representative G.G. Bramhill informed members of the Lambton Farmers' Cooperative Society that cooperation would increase their profits, since co-ops provided a way to cut many unnecessary expenses.[38] Both officials focused on marketing co-ops and referred to them as a means to financial gain, ignoring the many other benefits that arose out of such activity.

In early 1915 the society, although not affiliated with the UFO, sent two delegates to the UFO annual convention to learn about the new movement. One of the delegates, Peter Gardiner, "was not allowed to escape until he was placed on the directorate of the United Farmers'

Association." Gardiner was a member of the Osborne Club (a subgroup of the society), and he provided some insights into its operation. Most of the club's business was done through local merchants, and it appears that the chief business was that of purchasing supplies, mostly in ten- or twenty-ton orders. If a farmer paid for the goods upon receipt, one-half of one percent was added to the price to cover expenses. If the goods were not paid for immediately, a further one percent was added after thirty days. Gardiner also mentioned that a farmers' club had been established in nearly all of the forty school sections in Lambton, a not unreasonable claim, given that there were over thirty clubs by mid-1915, and thirty-seven clubs by January 1916. In addition, although the aim was to build a strong organization within the county, the members were beginning to see the benefits of combining with an already strong organization, such as the UFO. In addition to cooperative buying, the members of the society were attempting to arrange to sell their produce directly to wholesalers. Although relatively small in size, the society seems to have had success in its first few years of operation.[39]

In early 1916 a meeting of the society was convened in Petrolia. The meeting was to have been held in the agricultural office, but it was moved to the larger Victoria Hall to accomodate the crowd. The issue at hand was the need for all members to put up a ten-dollar note to be used as "collateral to form a basis of capitalization." The motion to this effect was passed almost unanimously. Interestingly, the society's leadership mentioned that the giving of a note was a personal act, that members were not bound by it, and that they could withdraw at any time without other farmers knowing about it.[40] Later that year the society asked members to put up another note, this time for twenty-five dollars, and this too was agreed to. Local bankers had advised the leadership that they would lend the society up to 150 percent of the notes' face value. Evidently business had picked up, and in December 1916 one club reported doing some $11,000 in business in the past four months, "thus saving hundreds of dollars to themselves in the turnover."[41]

Shortly after, the Forest United Farmers Association, which was affiliated with the UFO, emerged from and replaced the Lambton Farmers' Cooperative Society. Its directorship consisted of the president and secretary from each of its member clubs. The association hired Anson Groh – who had been dismissed by the UFCC for incompetence – as its manager.[42] Its object was to "secure more efficient cooperation among the various clubs, so that they can buy in car lots, save in the distributing charges, etc." Soon after its inception, members realized substantial savings through collective buying. A car of seed corn was purchased for ninety-one cents per bushel, when the local dealer's price was $1.15, which meant a saving of $240 on a car of one thousand bushels. In July 1917 the association consisted of forty clubs and some three hundred members.[43]

Running the association was no small feat. The manager took orders from local secretaries, coordinated shipments, and collected money for purchases. The goods were sold for delivery right off the car, but provisions were made for those unable to pay by storing their goods for a nominal fee. The manager received a one and a half percent commission on all farm supplies, two percent for groceries, and five cents per bag for sugar. Interestingly, provision was also made for non-members' use of the association. Outsiders paid one percent more than members, but this policy was not enforced. According to A.E. Vance, the association's secretary, the enterprise "has been organized on the principle that by being generous to outsiders they will soon be secured as members."[44] In 1917 roughly $25,000 in business was undertaken, mostly in feeds and flour, although some groceries, salt, and sugar were also handled. In Groh's first report, "a good margin over expenditure was shown."[45] What is noteworthy about all of the association reports is that at no time was there any mention of assistance from the state; it seemed to be a purely locally directed concern.

Writing to the *Sun* in 1919, Vance elaborated upon some of the features of the organization. First, he noted that there were several clubs affiliated with the association, and that the clubs normally consisted of no more than forty members each. This was a deliberate policy, as the smaller clubs provided a "means of social and educational as well as business advantages – a condition which does not exist to such an extent where clubs have a large membership." Although the plan was working well, Vance felt that "to be a live young concern a manager should be engaged permanently on a salary basis, warehouse facilities should be provided and a stock of all kinds of feeds should be kept on hand at all times." The association remained profitable but did not abandon the idealism of cooperation. As late as 1920 it ran advertisements in local newspapers with messages such as "To eliminate the Capitalist and Profiteer is Simple! Accept his responsibility and carry out his risks and retain for yourself his gain by cooperating." Financial gains could be made, but features such as accepting responsibility, defeating the profiteer, and gaining self-confidence were equally important.[46]

Simcoe County also had very active cooperative organizations. Large amounts of wool, for example, were being marketed cooperatively by 1918. In addition, the Orillia Cooperative Shippers sold five hundred hogs in 1918 with receipts of nearly $17,000; and although it had been formed only in July 1918, the Union Cooperative Shippers of Collingwood could boast of shipping $16,000 worth of hogs, cattle, and sheep by year's end.[47] Nine months after it was formed in 1917, the Ivy and Thornton Stock and Grain Company had one hundred members and was capitalized at $100,000. Soon after its establishment this co-op took over an elevator in Thornton and, in its first year of operation, sold over $210,000 worth of grain and livestock for local farmers.[48] Co-ops in Nottawa, Kirkville, Valley, and Batteau, all small clubs, agreed to a common shipping day so as to ensure full cars – a fairly common practice throughout the province.[49]

The early success of cooperation in Simcoe is clear from letters sent to the *Sun*. In 1919 a farmer from Stayner, who referred to himself as "Sunshine," wrote: "Our club seems to be a success if you can judge by the broad smile the farmers wear for a few days after they have shipped a car load of live stock." Another Simcoe correspondent, "G.W.H.," wrote: "Our live stock keep travelling to the city. Two and three cars per week and our shipments ran pretty near seventy thousand dollars since March ... It is up to every township to organize as unity means strength."[50] By 1919 there were at least fifty-nine UFO clubs in Simcoe, each making regular cooperative shipments.[51]

UFCC leaders often visited local clubs to spread the gospel of cooperation. J.J. Morrison and John Kennedy, Vice President of the Grain Growers Grain Company (GGGC), addressed a meeting of farmers' clubs in Simcoe in 1916. Under the existing system, Morrison contended, the farmer had no input with respect to the price of his produce: "He took eggs to the local store, and the dealer counted the eggs and fixed the price." By cooperating "the consumer and producer would be brought together," middlemen would be eliminated, and the farmer would have some agency in determining the final price of his produce.[52]

Kennedy and Morrison returned to Simcoe the following year, this time accompanied by E.C. Drury. After discussing the merits of cooperative marketing, Kennedy said he was not in favour of cooperative purchases of groceries. He believed that only those articles that could be bought in car lots, such as machinery and sugar, should be purchased in such a manner because he did not want to interfere "with the merchant's trade in shelf goods."[53] The state's bias against consumer co-ops was beginning, it seems, to find a voice amongst the UFCC leadership.

In Lanark too there was a considerable amount of cooperation before the UFO became a force in the county. The first egg circles, for instance, were formed in 1914. The Ramsay Farmers' Club reported good years in terms of cooperative selling in 1914 and 1915, and at times it devoted entire meetings to the subject of cooperative buying and selling.[54]

The club also brought in speakers, such as the manager of the nearby Lansdowne Cooperative Society, identified only as Mr Webster. His society was primarily a marketing concern, buying only seed and some supplies cooperatively, and much of his speech centred around the advantages of co-op marketing. He gave advice, however, that was relevant to all cooperators: "The first thing in cooperation is organization. Organize thoroughly, so thoroughly you will be able to have something to say about prices you will pay and prices you will receive."[55]

In 1915 the club was visited by F.C. Elford, a poultry specialist from the Central Experimental Farm, who spoke on the need for improved egg marketing,[56] and in late 1917 the club met to discuss the possibility of purchasing feed in car lots and of employing a manager to handle shipping and distribution. The local agricultural representative, Fred Forsyth, attended the meeting and took part in the discussion. Many of the farmers argued that money could be saved by eliminating middlemen and by dealing directly with wholesalers. A contrary opinion was expressed, probably by Forsyth, that employing a manager merely set up another middleman. In addition, the club had to think of other expenses that would be incurred, such as interest, insurance, and storage of unclaimed goods. In the end, the members decided to study the matter further.[57]

It appears that those advocating increased cooperation eventually won out, and in February 1919 the club decided to incorporate. At the same time, the question of affiliation with the UFO was raised, and club members agreed to proceed. According to a newspaper account, "It might not be possible to buy any cheaper than at present, but it would give the club a standing that it did not now possess. It was a step in the direction of unifying the farmers and bringing them more closely together. In the past they had been a football between the political parties, and instead of acting in unity very often one farmer nullified the action of another."[58] Affiliation with the UFO, then, represented more than increased returns. Political issues were at stake, and from an idealistic perspective, a closer connection to the UFO would serve to unite members in a common cause.

The formal act of affiliation came the following month during a meeting at which A.A. Powers, head of the Farmers' Publishing Company, was a guest speaker. Powers referred to past attempts at organization, such as the Grange and the Patrons of Industry, and argued that they failed because, unlike the UFO, they had "no centre or head." Powers asserted that farmers should "demand from Capital recognition as helpers in wealth production ... Why is price fixing applied to the farmers' products any more than to manufacturers' profits? Because manufacturers are organized and influence governments, and farmers are not."[59]

In early 1917 a meeting of the Perth UFO was addressed not only by provincial president R.H. Halbert but also by a Mr Marcellus of the Live Stock Branch of the federal Department of Agriculture. Marcellus, who spoke first, extolled the virtues of cooperative marketing. In particular, he urged the formation of egg circles, which in his view enabled farmers to establish different grades of eggs and allowed for a steady flow of product to market. Moreover, as he told his audience: "In forming an egg circle you manage it entirely yourself; we take no financial responsibility. Mr Forsyth, your district representative, or myself, will be willing to lend all help possible. We recommend it highly as a benefit to you." George Noonan, president of the Perth UFO, told Marcellus that an egg circle had been considered for some time, and that his advice would likely be acted upon. Marcellus then provided the club with a set of by-laws that were "in use wherever egg circles had been formed."[60] It is not known whether the members adhered to these by-laws or whether they instead took Marcellus at his word and organized the circle to their own liking.

Halbert spoke next. He emphasized that farmers had to keep production levels high, but that they should also use care in keeping track of where their products went, "watching that the iron grip of organized capital can be effectively dealt with." The solution, naturally, was cooperation. After pointing out the benefits of such a course, Halbert "cautioned the local branch to beware of smooth efforts of other lines of organized trade to try and take business away from the UFO by offering supposed inducements."[61]

One of the final speakers, Reverend A.H. Scott, asked Halbert why the UFO opposed the Cooperative Bill that was being debated in the Ontario Legislature. Halbert's response is worth quoting at length:

> Mr Halbert said the bill affected their organization on account of the government wanting each club ... to form a county organization. This was not necessary. The UFO was opposed to creating such a middle organization as the government proposed ... The Bill also provides that the business of the local branches of the cooperative company shall be carried on under the supervision of government officials. This was quite unnecessary, *as the local branches were quite capable of managing their own affairs without outside help.*[62]

Evidently, the UFCC and its increasingly top-down approach was not to be considered outside help.

The rest of Lanark was also growing in terms of cooperative strength. By April 1919 the recently organized Carleton Place club had roughly two hundred members, the Perth club was strong, and the Smiths Falls club was being reorganized to include the nearby Elmsley Club.[63] According to Dan Hogan, county director, all of Lanark would be organized by mid-summer, and by June clubs were established in Lavant, Darling, Dalhousie, and Lanark townships. In August 1919, at a UFO meeting in Smiths Falls "Prices were considered upon cars of coal, shorts, screenings, flour and feed, and it was decided, in view of the excellent quality of the last consignments ... to order several more through the Cooperative Company."[64] Simply put, the cooperative spirit was alive in Lanark.

The growth and popularity of the cooperative side of the movement was evident in June 1919 when R.W.E. Burnaby, president of the UFCC, visited Smiths Falls to address the club on the possibility of establishing a cooperative store in that town. Club members eagerly agreed to proceed, and plans for the establishment of the store took shape during August. A building was secured, and a reported three thousand dollars in capital, which was to be used to stock the store and to take care of other expenses, was deposited in a local bank. Cecil Hitchcock, a former Smiths Falls merchant, was hired as manager. By late October the store was open, and its first advertisement announced that it was "a store at which anybody may buy anything we have and all may buy at the same price."[65]

As with other similar UFO enterprises, little remains in terms of documentation regarding the Smiths Falls store's financial fortunes. However, while speaking to the Smiths Falls UFO in late 1920, A.A. Powers briefly described its fiscal state. By that time the store had $1,362 in local capital stock. Aggregate sales for the first year came to $71,366, and gross profits were $8,563. As of 31 October 1920 the store had goods valued at $12,076 on hand (selling price). Expenses for the year amounted to $7,743, which left a net profit of some $800. Powers reported that all seemed well but frequently referred to the lack of subscriptions to local stock. To illustrate his point he noted that the nearby Kemptville store had over ten thousand dollars in local capital stock. Powers's concern was that the central UFO had to use its own capital to make up the difference between the value of goods in the stores and the amount subscribed to locally.[66]

Despite promising beginnings, UFO stores began closing throughout the province in 1922. Speaking at the UFCC convention in December of that year, Burnaby blamed the members for not patronizing them and noted that farmers "still had a good deal to learn about cooperation, and that there had been too many of the Judas Stripe even in official positions."[67] Burnaby then provided other reasons to account for the failure of the stores: the UFCC's inability to extend credit to consumers (unlike the competition, which could); the loss of buying power among farmers owing to postwar conditions; and the difficulty of managing stores at such great distances from headquarters.[68] The final two stores closed officially in November 1923, although some (such as the one in Forest) remained in business, but with no formal affiliation with the UFCC.[69]

LOCAL COOPERATION AFTER THE 1919 PROVINCIAL ELECTION

After the 1919 provincial election, all signs pointed to a continued, if not increased, support for local cooperatives. This was certainly the case in Lanark County.[70] Sales at the Smiths Falls and the Perth stores continued to be good (see Table 19), and even the relatively modest Glen Tay Club conducted $10,000 in business in 1920.[71] And in early 1920 the United Dairymen's Cooperative Company (a department of the UFCC) set up a branch in Middleville.

The Carleton Place club had a membership of 155 by late 1920 and averaged over seventy members per meeting in 1921 and 1922.[72] Business was booming for the Smiths Falls club. At its annual picnic in 1920, club president John Willoughby boasted that "one year ago this month the club brought in the first carload of feed, and today could show a record of over $130,000 worth of business going through the Bank of Commerce, without saying anything of livestock. The club had also bought a large flour mill and had an up-to-date store." Willoughby also made an important connection in his speech that often went unmentioned by government, and even UFCC, officials: "There had been a feeling of danger in some districts that when we had cooperative stores and livestock shipping that it would be considered all which was necessary and there was no need to keep up the clubs. But that was getting away from the ideals and principles of the UFO altogether. We must keep up the clubs in order to progress along the ideals and principles of the organization."[73] Profits were fine, but members had to ensure that they did not lose sight of the larger picture. Not content to rest on past successes, the club planned for an even better year in 1921. Early that year each member was asked to pay five dollars to purchase and install scales near the town's stockyards.[74]

One of the most dramatic cooperative episodes in Lanark occurred in March 1920, when Smiths Falls-area UFO members purchased the Woods Mill in that town (and renamed it the Rideau Milling Company), apparently without consulting the central office. Although the

Table 19
Sales for the Smiths Falls and Perth UFO Stores to October 1920

SMITHS FALLS	
November 1919	$ 3,858.61
December 1919	4,501.78
January 1920	5,266.04
February 1920	4,779.50
March 1920	8,255.32
April 1920	9,514.93
May 1920	10,292.06
June 1920	7,072.02
July 1920	6,498.43
August 1920	5,853.88
September 1920	7,337.57
October 1920	3,828.39
	$67,609.02
PERTH	
July 1920	$ 2,199.87
August 1920	2,765.61
September 1920	4,294.32
October 1920	3,860.64
	$13,120.44

Source: University of Guelph Library, Leonard Harmanuco Collection, XA1 MS A126037

exact price was unknown, it was rumoured to be roughly thirty thousand dollars. Later, it was revealed that the company was capitalized at forty thousand dollars, which had been raised locally through the sale of twenty-five-dollar shares. Evidently, no farmer invested more than one thousand dollars.[75]

Not long after the purchase the executive of the UFCC decided to lease the mill from the local farmers "at a rate of eight percent clear upon their investment." The UFCC wished to set up a milling department, and the Smiths Falls mill, if properly managed, had considerable potential. One UFCC official, T.P. Loblaw (who later established a chain of grocery stores in Ontario), was sent to Smiths Falls, where the local executive told him that the UFCC could lease the mill for $3,500 per year plus taxes. Loblaw's account of what transpired bears the hallmarks of a less-than-honourable deal: "Knowing that the Board of Directors were under the impression that the offer was $3,500, I agreed with them at eight percent clear on purchase price. This would be less than $3,500 per year."[76] The UFCC, not the local club, appointed the miller, T.H. Squire of Madoc, at a salary of forty dollars per week, and he was instructed to "look after all the details" and make improvements on behalf of the central office.[77]

The problem of having a miller who was unfamiliar with the community he served soon came to the fore. Shortly after the mill began operating, a farmer at a local UFO meeting accused the GGGC of depriving the mill of grain, a charge that Squire denied, saying that if area farmers would grow enough wheat then the mill would run full time. Another farmer observed that Lanark was not good for wheat growing, to which Squire replied that, if the land was worked the way the farmers' fathers had worked it, then wheat could be grown in abundance. Yet another farmer claimed that the climate had changed since their fathers worked the land, a contention that was dismissed by Squire, who observed "that the same Being was always in control of the weather."[78]

By late 1920 it was discovered that the mill was in disastrous financial shape. The UFCC commissioned auditor A.F. Low to attempt to find out what had happened, and the report he produced, entitled "The Smiths Falls' Haystack," placed the blame squarely on Squire's shoulders: "The cash book and sales book which I instructed [Squire] to start, last visit, were discontinued as soon as my back was turned ... Now, one of the many baseless notions, which Mr Squire entertains, is that he is absolute at Smiths Falls, subject only to his own personal inclination – a kind of modern despot as far as I can make out."[79] According to Low, it would take countless hours to sift through Squire's scraps of paper to determine the real state of affairs, and the final results would be distressing.

Low's prediction proved accurate, and by February 1921 the situation had become so hopeless that the possibility of converting the mill into a cold-storage plant was considered.[80] The *Record-News* thought that it might be a wise move, especially since the federal government, recognizing the value of municipal cold-storage plants, had announced that it would contribute up to thirty percent of the establishment costs. The plans to convert the mill were not realized, and in November 1922 the UFCC directors decided to dispose of it.[81] In February 1923 the directors agreed to instruct the manager of the Grain and Feed Department "to take steps to have a UFO brand of Feed and Flour put on the market and that he have authority to negotiate with any *reliable* milling company" to attain this end.[82] The mill ceased operating on 1 June 1923, leaving Smiths Falls UFO members smarting from the failure of their venture.[83]

It should be stressed that the purchase of the flour mill in Smiths Falls was not an anomaly. In other districts several UFO clubs acted, or attempted to act, in a similar fashion. In August 1920 the various farmers' clubs in West Rama Township secured an option on the Brechin Flour Mill. The clubs endeavoured to raise the necessary capital by selling stock at one hundred dollars per share. It appears that insufficient capital was raised to purchase the mill; but in October of that year the clubs purchased the Harris Brothers Flour Mill in Brechin.[84] In Lambton, the Thedford UFO club purchased a church in Bosanquet Township for five hundred dollars in 1920 and converted it into a storage building. In addition, UFO members from the Inwood area purchased the elevators at Inwood, capitalized at twenty thousand dollars, raising the required funds by selling twenty-five-dollar shares.[87]

LOCAL COOPERATIVES, THE UFCC, AND THE STATE

During the years in which UFO-affiliated cooperatives were thriving in Lanark, the state was active as well.[86] In 1920 F.H. Buker, of the federal Department of Agriculture, approached the Smiths Falls UFO club with the intention of adding it to the countywide network of egg circles, tied together by a central organization. According to the local press, after Buker explained to Smiths Falls farmers the advantages that would accrue to them through the formation of such a system, he then revealed the government's motivation: "Where the Government was interested in the matter was in the raising of the standard of Canadian eggs. These were even now a prime favourite upon the British market and the cooperative movement was another step in the direction of making the Canadian product more popular and reliable with the result that they would bring a higher price than the eggs of other countries where no attempt was made to grade them, according to quality."[87] The company that emerged from this endeavour, the Lanark Cooperative, Limited, was formed in November 1919. At the inaugural meeting in Perth, the district agricultural representative "was in charge of the details ... and had carefully arranged everything." By 1920 it had 1,800 contributors and 112 members, each of whom gave a promissory note for fifty dollars.[88] By July of that year it was doing roughly four thousand dollars' worth of business per week. According to the *Ottawa Farm Journal*, the organization was the only one of its kind in Canada "to grade and sell all its eggs by standard and subject to Government inspection." The co-op was so large and financially strong, the paper noted, that "many farmers who are not members are deriving benefit from the organization because the cooperative practically sets the store price in Perth."[89] Despite being an impressive organization (one of the largest of its kind in Canada), it was surprisingly short-lived; it went into liquidation in 1922. The local agricultural representative, who closely monitored the progress of the company, blamed the failure on poor management. In the 1921–22 report, the representative wrote cryptically that "while it is a hard blow on the local Representative, it is encouraging to know that he is not held responsible for the failure ... by the shareholders."[90] The UFCC set up a similar venture in 1924, with similar results.[91] In the end, the most successful circles were the ones that were purely local concerns.[92]

Regarding dairying, in 1922 Lanark farmers were informed of the intention to form the Ontario Cooperative Dairy Products Company Ltd. One of the organizers, a Professor Colquette of the Ontario Agricultural College (OAC), was on hand to explain the workings of the new firm, which had as its initial goal the securing of fifty percent of the province's dairy market. Factories from all over Ontario were to sign binding contracts for a three-year period and would then be represented on the central board. According to press accounts, it was planned "to market the product in a manner that will eliminate gluts on the market and with the aid of cold storage warehouses spread the marketing season over the full year."[93] A few farmers expressed scepticism, but ex-UFO MPP Hiram McCreary warned them that the industry was "facing a total collapse unless something was done,"[94] and that the company seemed to be the best way to remedy the situation. By January 1923 the company was ready to conduct business, although only on a modest scale and with limited success.[95] Later, both the Ontario and federal governments assisted Lanark in the construction of a cold-storage facility in Perth.[96]

In Lambton, cooperation proved quite popular among UFO members. Members of the Wanstead branch, formed in 1921, realized by 1922 that their club was too small to ship every two weeks without getting assistance from non-members. As a result, the club devised a scheme, based on one proposed by Aaron Sapiro, to have farmers sign contracts to deliver hogs at certain times, thus ensuring full cars and lower rates. There was another reason for implementing such a scheme: by using a contract system it would prevent some farmers from using the club "as a lever to get a higher price from the drover."[95]

In mid-1922 R.J. McMillan, a UFCC director, asserted at a Wanstead Club meeting that area farmers were receiving two dollars to three dollars more per hundred for their hogs than they had previously. He also noted that twine was being purchased at some four cents less per pound because it was being bought cooperatively. McMillan then told his audience that "the farmer has always been an individualist, both in production and marketing. The system under which he marketed his products made him a competitor of every other farmer with the result that he played directly into the hands of an army of non-producers, the middleman, who in almost every instance made a larger profit than the producer."[96]

Business improved for the club, and it enjoyed steady growth. Not only was it making regular shipments to stockyards by 1924 (with an annual turnover of eighty thousand dollars) but it was also engaged in bulk purchases of supplies. In fact, business was so good for the club that it received letters patent in 1924 and formed a joint stock company known as the Wanstead Farmers' Cooperative Company, Ltd. Upon its incorporation, a correspondent wrote that "it now remains for the club to give attention to other lines of endeavour, educational and social, which were the primary objects in the formation of the United Farmers."[97]

The Forest United Farmers' Cooperative Association also continued to do well. The turnover for the association in 1918 had been $60,000; in 1919, $135,000; and in 1920, $323,000. Stock shipments were being made weekly, and, as mentioned earlier, a UFO grocery store was established in Forest in late 1920.[98]

By 1924 a UFO egg circle had been established, and members in Lambton had the option of having their eggs sent to Toronto to be pooled or having them sold locally at current market prices. The arrangement appears to have been the result of local initiative, as it was not found elsewhere. As late as 1926 the circle was expanding in terms of the amount of business it conducted.[99]

In 1927 the Forest United Farmers' Cooperative Association was approached by the UFCC with the intention of setting up a wheat pool. Manitoba Wheat Pool president Colin Burnell addressed a meeting of the association and explained how the contract system worked in Manitoba. Under the terms of the contract, farmers committed themselves to selling exclusively to the UFCC for a period of five years. Contracts were then distributed to members, who were asked to examine them and vote on the advisability of forming a pool. A number of farmers did sign contracts at the meeting.[100] Aside from activity related to wheat, 1927 was also a reasonably good year in general for the co-op; it shipped 1,260 hogs, 260 sheep, 77 calves, and 70 cattle. The net proceeds for the year amounted to $32,343.[101]

As in Lanark, the state was quite active in Lambton during the 1920s. It emphasized celery and other cash crops, in addition to efforts aimed at increasing livestock and dairy production. Celery was seen as a particularly valuable crop in Lambton, which had some of the best soil in Canada for growing it. Because celery required a special system of refrigeration, it was felt that the lack of proper facilities prevented Lambton farmers from attaining full production. Consequently, in conjunction with the federal and provincial governments, the Lambton Growers' Cold Storage Company was formed in 1933. The total cost in erecting the cold-storage plant was $15,717.12, and although common stock was sold and a bank loan secured, it appears that the plant would not have been built had the company not received a $4,500 loan from the provincial government.[102]

There is evidence of renewed confidence and activity following the 1919 election in Simcoe County as well. The Sunnidale Club, with 105 members, joined forces with the Nottawasaga Club to ship cooperatively out of a jointly owned stockyard at Stayner. Up to three cars per week were leaving for Toronto by 1920. That same year a UFO store opened in Barrie.[103] In 1921 Phelpston held its first annual meeting of the UFO Shipping Association, which was headquartered near Barrie. Later that year the club purchased weigh scales for Phelpston. Even something as innocuous as this managed to draw scathing comments from the urban press: "It is the wish of your scribe that they through usage come to realize the accuracy of said scales, and to find out the chronic kickers who always felt that the scales should weigh their livestock according to their imagination."[104]

Despite the support of many Ontario farmers, the UFCC began to experience serious losses during the early 1920s from which it was never able to recover. In order to renew interest in the company, a number of schemes were undertaken. For instance, in early 1923 UFCC representatives travelled the province to encourage farmers to form UFO egg circles, and two officials, E.C. Drury and C.E. Merkely, visited Simcoe. They explained that under the scheme the company graded members' eggs and then marketed them as demand warranted. Expenses such as buying crates and shipping charges were to be borne by the individual circles, and a fee of one cent per dozen was charged by the company to market the eggs. In effect, according to Drury, "instead of selling to the wholesaler, who in turn makes a profit, the farmers became the wholesalers themselves." Merkely had an even stronger message for those in attendance: "It is pretty hard to come back with any cooperative ideas to the farmers after the losses they have sustained ... The losses ... may have been due to bad business ventures. You must not sit back, for you are partly to blame in putting these men there."[105] Despite being berated by a leadership increasingly focused on profit making as opposed to the observance of egalitarian principles, almost everyone in the hall agreed to join the circle.

Even with UFCC manipulation, the cooperative spirit remained very much alive among Simcoe farmers. In 1924 the Stayner club bought out a drover's business and stockyard and began shipping soon after. In 1925 it was reported that 107 cars of livestock had been shipped that year, at a value of $153,000. Some 29,538 bushels of wheat, 3,451 bushels of barley, and 14,000 bushels of oats had also been shipped; and one car of salt, six tons of twine, and seed corn had been purchased cooperatively. In 1926 the club incorporated as the Stayner Farmers' Cooperative Company (which also included the Sunnidale Corners Farmers' Club). In 1926 an elevator was constructed and shipping revenue amounted to $227,686.[106] In 1927 the Nottawasaga club boasted of shipping $103,957 of produce, of which $96,882 was distributed among the farmers who shipped with the club. Moreover, in 1929 farmers in the Barrie area formed the First Cooperative Packers of Ontario Ltd, a co-op independent of the UFO and run on the Rochdale principles.[107]

In Simcoe the state was active as well, primarily in the areas of wool, fruit, and livestock. The impact of its efforts was positive in some cases and ambivalent in others. With the assistance of the provincial government, several initiatives were undertaken to encourage cooperative marketing.[108] Wool growers in Simcoe, as elsewhere, were supplied with twine and sacks free of charge if they shipped their wool to the Ontario Sheep Breeders' Association in Guelph.[109] With respect to livestock, the state – through the Board of Agriculture – attempted to work with local clubs, many of which were UFO clubs, to set up a countywide shipping network based on clubs entering into agreements with local shipping agents. A few individuals and clubs did sign contracts, but the fifteen clauses in the agreements, which effectively bestowed power in the agents, probably made most farmers balk at such an arrangement.[110]

UFO COOPERATIVES BY THE MID-1920S

Despite all of the efforts of individual UFO clubs, after 1925 cooperation through the UFO entered into a period of decline.[111] Although the reasons for this are numerous, it can be argued that one of the primary reasons was that the central UFO made many mistakes, both in financial terms and in its emphasis on certain aspects of cooperation, especially cooperative marketing.

The latter case is well illustrated in the correspondence course on cooperation offered by the UFO Educational Department for 1924–25. In the first lesson, students were told that the purpose of cooperation was to "obtain for the farmer the advantage of large-scale organization, so that by collective buying of farm supplies and the cooperative marketing of farm products, all unnecessary middlemen may be eliminated." The idealistic benefits of cooperation were relegated to a few short sentences in the section describing the Rochdale Pioneers.

The second lesson in cooperation pointed to the success of Danish farmers and their cooperatives, which were "in part supported by the state." The three subsequent lessons were all entitled "Cooperative Marketing." In these lessons, the virtues of commodity-plan co-op marketing, of pooling, of contracts, and of pleasing consumers were extolled. In summing up the future of the UFCC, students were told that, "in line with the commodity marketing idea, the policy of the company is to develop the business as a big central organization divided into separate departments each handling a separate commodity or one or two closely allied commodities. *Each department will stand on its own feet and the profits of each department will go to the patrons of that particular department in proportion to the volume of business they contribute to it.*"[112] No discussion about the values of cooperation to a community, the benefits of local control, or virtually anything that was being discussed a mere five years earlier. What was important now was that farmers maximize the return on their investment, and that they have control over the sale of their products.[113] Such were the lessons that young UFO members were to learn.

There is ample evidence that the UFCC concentrated on profitability rather than the intangible features of cooperation. For instance, UFCC-sponsored speakers told rank-and-file members that they should behave more like the Canadian Manufacturers' Association (CMA) and attempt to control output and prices. For an organization that vilified the CMA, this was a rather bizarre comparison. Others spoke on the need to manipulate consumers into buying items they did not need.[114] In addition, by the mid-1920s the UFCC supplied club secretaries with confidential price lists that enabled the secretaries to pocket a percentage of the fees of a transaction. This tactic resembled what the UFCC had initially fought against – the bribing of local members to do business with other firms. This time, however, the bribery was effected internally within a farmers' enterprise.[115] By adopting these tactics the UFCC began to act like the capitalist firms it had originally condemned. Such behaviour could not help but have an effect on members.

CONCLUSION

The foregoing raises questions related to the influence of the state and of the central leadership of the UFCC on the cooperative efforts of local farmers. First, in serving the interests of capital, the state acted in ways that adversely affected these efforts. Yet in objective terms, farmers are capitalists. They invest in land, seed, stock, and implements and, at times, employ labour in the hope of obtaining a good return on their investment. If this is so, then why did the state not work on their behalf?

The state did in fact appeal to and attempt to bolster the capitalist impulses of UFO members. The message that was constantly conveyed – that cooperative marketing meant greater returns for farmers – amply demonstrates the efforts on the part of both the federal and provincial governments to encourage farmers to think in terms of profit making. But did UFO members consistently see themselves as capitalists? As noted elsewhere in this study, there was a strong contingent of idealists in the movement, individuals who saw cooperatives as more than money-making ventures. There were financial gains to be made from cooperating, but even with respect to "profits" the benefits farmers hoped to realize were not so much actual monetary rewards as they were fair returns for their labour. In short, the "value" cooperation added was its elimination of the "robbery" farmers endured from middlemen who did not produce food but still reaped the greater returns from its sale. In this sense, despite being objective capitalists, UFO members were subjective victims. They saw themselves as hapless slaves of the Big Interests, with no agency regarding how much they received for their labour. The Big Interests dictated what price farmers paid for implements and what price they received for produce. In this context, rank-and-file UFO members believed that they were victims who were denied a "square deal." The purpose of cooperation, then, was to facilitate their breaking free from this bondage. When one adds other idealistic notions the rank and file held (such as participatory democracy, equality, and decentralization), then one obtains an idea of why cooperation, with its focus on mutuality and grassroots democracy, held such appeal for rank-and-file members.

This leads to the second question. If it is true that many UFO adherents were drawn to the idealistic benefits of cooperation, then why did the Drury government not act to instill these values in cooperatives while it was in power? As noted in previous chapters, there were moments when the aspirations of the movement's leadership did not always harmonize with the rank and file. In addition, UFO MPPs tended to act moderately once elected. To some extent, this may have been because they did not want to be seen as favouring one group over another (recognizing that they would need to appeal to more than farmers if they wished to be re-elected), but it is equally plausible to suggest that, given their inexperience in governing, they chose to be uncontroversial or were mystified by the responsibilities of administering an entire province.

In addition, one must keep in mind that the state does not consist solely of elected officials. It also includes the bureaucracy, judiciary, and educational institutions. The bureaucracy (Department of Agriculture officials and OAC faculty) concentrated on marketing cooperatives prior to the 1919 election. Thus, even if a UFO MPP wished to encourage the less-tangible elements of cooperation, he may have encountered a state apparatus that made no provision for or even acted contrary to his intentions.

Writing about cooperatives on the Prairies, David Laycock argued that "cooperative enterprise thus provided its participants with a sense of achievement and self-respect which is the *sine qua non* of transformative democratic action. Once the initial hurdles of social isolation and deference to the 'received order' had been cleared, prairie cooperators and their acquaintances could be recruited to a variety of political endeavours."[116] In light of what has been discussed in this chapter, the same could be said about farmers in Lambton, Simcoe, and Lanark. The problem, however, was that these members did not exist in isolation; various influences – the state, changing consumer demands, and even the UFCC – undoubtedly tempered their response to the economic injustices that they perceived. In fact, after the UFO formed the government in 1919, farmers may have felt confident that all levels of government would act upon their wishes. This may explain, in part, why there was a tendency to opt for centralized cooperatives rather than small, local ones.

If a group of farmers wished to form a cooperative, a dizzying array of options immediately opened up for it. The group could seek state support, affiliate with the UFCC, or maintain local control. Each choice carried potential benefits and potential pitfalls. If the group sought state assistance, the cooperative might, thanks to government backing, enjoy a strong, stable foundation upon which to build. Good returns on the group's investment might then accrue, with little liability. However, as a government-sponsored body there would be rules to follow, no partisan politics would be allowed, and a great deal of autonomy would be sacrificed.

If the group affiliated itself with the UFCC, it would be associated with a large, well-capitalized business. In addition, all members theoretically had a voice in the affairs of the company, and one might eventually become a director. Affiliation with the UFCC, however, meant giving up considerable autonomy. At the very least it carried with it the risk of being castigated for stupidity and bad business sense by the company's leadership, while in the worst case, it presented the possibility of large-scale business failures. In either case, it eventually meant that one's cooperative was affiliated with an enterprise that stressed the profitability of its ventures, making scant mention of the other, less-tangible aspects of cooperation.

If the group chose to go it alone, it would enjoy complete autonomy and could adapt to suit local needs and conditions. Equally important, if kept small enough it could be a model of democracy. Maintaining local control, however, meant that financial institutions might be unwilling to offer assistance, and that any problems that arose might have to be faced alone. Moreover, even a small financial setback could spell ruin.

Decisions, important decisions, had to be made.[117] In the end, it turned out that many purely local cooperatives, and those with the greatest measure of local autonomy, had a good chance at being successful, at least in terms of maintaining the ideals of the UFO. But UFO members in Lambton, Simcoe, and Lanark, not having the benefit of foreknowledge, often chose to let the state or the UFCC assist them in their cooperative affairs. And although these ventures were reasonably successful financially, other elements were lost that had serious consequences for the movement.

7 Conclusion

> Democracy ... isn't some ultraprogressive myth of a superbenevolent World As Should Be. The meteoric burlesk melodrama of democracy is a struggle between society and the individual over an ideal – a struggle from which, again and again and again, emerges one stupendous fact; namely, that the ideal of democracy fulfils herself only if, and whenever, society fails to suppress the individual.
>
> E.E. Cummings[1]

> The people believed that war would secure relief from autocracy, but autocracy cannot be slain by wars: they only force it to hide for a time after which it will return in a new form. The only sure way is for the people to dream of democracy, plan for it, and then secure it by legislation.
>
> Emma Griesbach[2]

In the foregoing account of the UFO in Lambton, Simcoe, and Lanark counties, I have described and analyzed the context in which the movement emerged and the composition of its membership. To reiterate, the three counties differed (in some cases considerably) in terms of soil quality, crops, and returns farmers received for their produce. Lambton farmers, situated in an area with good soil and a long growing season, fared reasonably well. In Simcoe, which had fairly good soil and a relatively long growing season, farmers employed mixed farming techniques and received reasonably good returns for their efforts. In Lanark, with its short growing season and poor soil conditions, farmers had considerably fewer options at their disposal. Since only certain crops could be grown with any success, many farmers chose to concentrate on livestock instead. Even then, the economic prospects in Lanark were not as favourable as they were in the other two counties under consideration.

Despite the different physical conditions and the varying degrees of prosperity in each jurisdiction, Lambton, Simcoe, and Lanark had large UFO memberships. In addition, as seen in the membership profiles, the UFO was an inclusive movement in all three counties, attracting members from every rank in rural society. Well-to-do farmers met and discussed issues with those who struggled to survive each year; people from diverse religious backgrounds fraternized during meetings; and young and old agrarians exchanged ideas. The evidence suggests that members were often neighbours. This is understandable: UFO meetings and social events provided an effective means for reducing the sense of isolation among rural residents, particularly farm women, and groups of people who knew one another would have taken advantage of the opportunity to assemble. As this suggests, the United Farmers had a marked impact on agrarian communities. Given the clusters of members in townships and their proximity to non-UFO members (as seen in the example of Simcoe), it is reasonable to assume that the UFO touched the lives of even more rural residents than raw membership figures indicate.

The UFO also largely mirrored the social composition of the counties, with two notable exceptions. First, the evidence suggests that the UFO was a young person's movement. Whether it was fear of conscription that initially prompted membership, or whether it was the idealistic premises upon which the UFO was built that attracted young people is uncertain. The fact remains, however, that age was one of the two variables that did not find an approximate match in the overall rural population.

Second, it appears that the UFO tended to draw more adherents of evangelical denominations than members of other religious groups. In Simcoe and Lanark (the two counties in which religious affiliation was listed in the assessment rolls), Presbyterians were overrepresented. Given the idealism of Social Gospel adherents, it is not surprising that one finds that they were a strong presence in the UFO. This does not mean, however, that supporters of other denominations were not members. Even if the movement did not exactly mirror the religious composition of the three counties, it remained a reasonably inclusive organization in that regard.

If all farmers were welcomed into the UFO, then the only difference between members and non-members was the idealism of the individual farmer. As the preceding chapters make clear, the movement did not want for idealists.

Not only did the three counties under consideration resemble one another in terms of the levels and character of the UFO membership but they also had an affinity of perspective – the concerns expressed by movement rank and file were, in the main, similar in Lambton, Simcoe, and Lanark. Despite the differences in farm size, wealth, and geographic location, there was a common response to external conditions. In all three counties the UFO was, at least initially, an idealistic organization that experimented with dissidence. This defining feature was demonstrated in its response to the Great War and the issues that arose from it.

Given the idealistic character of the UFO, it is understandable that it would be activist in focus and that this activism would find concrete expression in the 1919 provincial election and the 1921 federal election. During the years leading up to 1919 UFO members had lived through the greatest slaughter that humankind had hitherto witnessed, all – at least rhetorically – in the name of democracy. They also saw so-called patriotic businessmen reap huge profits from the carnage, while they themselves were being told by the state to produce more with fewer resources. Finally, adherents of the movement experienced the federal government's duplicity and cynicism first-hand with the revocation of conscription exemptions.

Throughout the province UFO members threw off the yoke of orthodoxy and began asking questions about the society in which they lived. Since they found many of the responses to their queries unsatisfactory, they formulated alternative approaches. In particular, they made an effort to check the influence of the anti-democratic Big Interests, the term UFO adherents used to describe the social class that possessed wealth and power sufficient to influence many aspects of life in Ontario.

Farmers believed that the Big Interests accomplished many of their objectives through political avenues. Time and time again agrarians saw legislators use a variety of means to protect domestic industry from outside competition. It was thus logical for UFO members to conclude that those who made the laws – the legislators from the old parties – were either in the back pocket of the Big Interests or were part of that class themselves. UFO members assumed that one effective means of checking the power of the moneyed interests was to nominate candidates from their ranks to challenge (and, it was hoped, defeat) competitors representing the old parties.

The political experiment was at first successful. Acting independently and spurred on by a few by-election victories, local UFO clubs fielded contestants in the 1919 provincial election. When the votes were counted, the UFO emerged with more seats than any other party. The 1921 federal contest yielded similar results. Having attained political power, many members eagerly anticipated the future. In their minds, the answer to Emma Griesbach's question – "Should Canada be a democracy, government of the people by the people, or should it be government of the people by a group and for the interests of a group?"[3] – was quite clear.

Despite these early successes, the UFO's venture into politics ultimately failed. The 1923 provincial campaign was a disaster, as were the 1925 and 1926 federal contests and all that followed. Owing to internal discord (particularly at the central level), the inability or unwillingness of UFO legislators to implement the reforms advocated by the rank and file, the resolve of the old political parties to fight the UFO with every tool at their disposal, and the power of hegemony, the political side of the movement was largely exhausted after 1926. The zeal for political action disappeared.

Accompanying the political defeats was an overall decline in UFO membership. Precise numbers *vis-à-vis* local adherents are difficult to obtain but provincial figures are available. Across Ontario membership declined from a peak of sixty thousand in 1920 to twenty thousand in 1926, and then to no more than fourteen thousand in 1929.[4] In Simcoe, where the minute books of three clubs survive, membership in these clubs peaked in the period 1919–20 and then rapidly diminished.[5]

Despite the efforts expended by members in contesting elections, the activism of the UFO was not confined to narrow political concerns. As a dissident movement it identified as problematical several dimensions of Ontario life that went beyond conventional party politics. For instance, the attainment of equal status for women formed an important component of the UFO's agenda. Initially women enjoyed equal status within the movement as a whole, and then they were accorded their own separate organization, the UFWO, which allowed them to explore more directly issues that had an impact on their lives. Led by individuals such as Emma Griesbach and Agnes Macphail, rank-and-file women responded enthusiastically at first. Due in no small part to Griesbach's "Sun's Sisters" page in the *Sun*, they began to identify and critique some of the societal values that relegated them to the status of second-class citizens. Letters offering wide-ranging opinions on a host of topics crowded the columns of the "Sister's Page," and clubs sprang up throughout the province. The UFWO offered many women their first experience of electoral politics and, by this means, certainly educated them in equity activism as well as in the expression of opinions on equal rights.

The UFWO, however, did not escape the dilemma of being auxiliary to a large, male-dominated organization. Gradually, as it was realized that equality in the movement was not as complete as it could be, many farm women became disillusioned. This demoralization was evident in the decision some took to restrict their memberships to associations that were gender-specific, such as the Women's Institute. In these organizations, as recent literature suggests, women worked over the long term to attain a measure of control in their communities, while also continuing to educate themselves. Though its active life was short, the UFWO was a significant body. It did not achieve full equality for women in the agrarian movement, but it gave women who had hitherto been afraid to state their opinions a voice, through the *Sun* or through local UFWO meetings.

The UFCC represented another attempt on the part of the UFO to engender change through unconventional means, in areas of Ontario life that fell outside of the purely political arena. UFO members became infused with the belief that cooperatives would allow them to initiate and conduct their own affairs; indeed, for a time it was commonly held that cooperatives would eliminate middlemen and others who unjustly profited from the farmer's labour, and that they would allow people to experiment with the locally based democratic structures that would ultimately transform society. This belief was short lived, however, and by 1930 the UFCC existed as little more than a state-approved marketing board.

The decline of the UFCC can be ascribed to several factors. Infighting, bad business decisions, increasing centralization, state influence, and hegemony all played roles in diminishing and ultimately defeating the company as an alternative mechanism for conducting commercial affairs. Although it remained a successful concern, at least in terms of securing good fiscal returns for its members, by the mid-1920s the UFCC was spent as an alternative force. By then it devoted virtually no time to impressing upon farmers that cooperation meant more than a few extra dollars in one's pocket.

The trials and tribulations of the UFCC provide a solid example of what UFO members faced, both from within the movement and from outside influences. Not only did the company have to attempt to counter the central precepts of a society in which capitalist concerns were considered the engine of economic development and the natural form of economic behaviour but it also had to contend with a state that bolstered these notions. Although the state did encourage certain UFCC activities, it only did so because these endeavours helped advance certain elements of the state's agenda, an agenda that differed markedly from that of UFO members. By the mid-1920s, there is little doubt that many farmers viewed the UFCC merely as a means to get produce to market and to receive a decent price for it. Although the enterprise did not, as a result, achieve its original goals, many agrarians who had access to the UFCC were able to experience the individual self-respect and the collective self-confidence that form the very heart of cooperation. Given the countervailing forces described in this book to alternative perspectives, it is understandable that experimentation with and adherence to provincewide cooperation was fleeting. Still, even after large-scale cooperatives such as the UFCC began resembling capitalist concerns, many local co-ops remained as an example of what the rank and file could accomplish, if given the opportunity.[6] Though they operated in the shadow of a large, centralized company whose leaders had accepted the notion that cooperation meant greater financial returns and little else, local co-ops continued to stress membership participation and the value of democratic, locally based decision making.

All this could be interpreted as hopelessly bleak. After all, despite the best of intentions, the movement attained few of the goals that it had set for itself. After the UFO disbanded in 1943, the Big Interests continued to operate much as they had before – the press, as a vital component of the old order, remained loyal to its masters; no new electoral systems replaced the first-past-the-post mechanism; and the old political parties emerged as shrewder organizations. Given this outcome, how could the UFO be seen as anything but a failure?

Yet in many important ways the movement can be seen just as readily as a success story.[7] The UFO had a profound impact at the local level. It came into being at a time when farmers' organizations, except for state-directed ones, were in disarray. Assuming responsibility, previously undertaken by the government, for uniting and educating farmers, the UFO played a vital social role, breaking down some of the isolation that farmers may have felt through the various events that it staged. Activities such as picnics, community entertainments, meals, sporting events, debates, and readings undoubtedly brought people together so that ideas could be exchanged and considered. The UFO went beyond the state, however, in its direction of such events. It added a political dimension and provided the venue for the expression and application of idealistic perspectives. UFO clubs enabled farmers to experiment with democratic forms and to join together to formulate probing critiques of their society and alternative ways of conducting social, political, and economic affairs. As seen throughout this study, local members exhibited striking energy, thought, and creativity.

If all this is so, then what happened? Why were members unable or unwilling to push the movement further so that a more democratic and equitable society could be attained? Why were they unable to advance their ideas to their logical conclusions? To be sure, the movement's progress was held back by internal struggles, an uneasy relationship with labour, a central organization that was more moderate than the periphery, the impatience of youth within the movement for change, and in some cases by ineptitude. But another factor impeded the movement even more severely.

To find a satisfactory answer to the question why the movement failed to fulfil its promise, one must appreciate the pervasiveness and influence of the old order in Ontario at the time. As we have seen, those who held sway over public opinion had many tools at their disposal. The mass-circulation newspapers, themselves corporations, presented readers with messages that encouraged them to support the *status quo*. Politicians from the old parties stressed the "naturalness" of the British parliamentary tradition, insisting that there was no better system of government in the world. At the same time, they emphasized the need to protect Canadian industries lest the country go bankrupt, or be faced with massive unemployment and poverty, or be overrun by the Americans. Those with an interest in preserving the *status quo* sought to discredit the UFO at every turn, arguing that it was a disloyal group containing seditious elements, that it was intent on installing a dictatorship of a minority over all Canadians, or that it was composed of misguided and naive idealists. From this perspective, it is remarkable that UFO members were able to break free to the extent that they did. Of course, the movement met such vehement opposition with good reason. UFO members mocked a parliamentary tradition thought by many to be sacred; looked irreverently at industrialists who supposedly represented the pinnacle of social and patriotic achievement; sneered at politicians representing parties that had legitimacy only because of their longevity; and derided a press that fabricated lies and distortions on behalf of the Big Interests. These societal elements had the benefit of "rightful status"; they existed, it was argued, because they represented the best that human beings could attain. The UFO was perceived as a significant threat because it presented a substantial challenge to all that was "legitimate." It had therefore to be countered and, ultimately, crushed.

In light of what has been written about agrarian movements in particular and about rural society in general, one can speculate on the question how UFO members perceived themselves. Did they see themselves as cranky Liberals in a hurry? Did they consider themselves to be independent commodity producers jealously guarding their *petit bourgeois* class position? Did they see their actions through the prism of family dynamics? What does this account of the UFO tell us about its members' self-perceptions?

It is misleading and far too simplistic to state that UFO members were merely impatient Liberals, or that they were *petit bourgeois* independent commodity producers who, *ipso facto*, were incapable of positing an alternative to the existing capitalist structure. It is inaccurate, as well, to suggest that they were paranoid cranks who saw a Big Interests conspiracy in every public policy and business decision. It is too patronizing, moreover, to argue that UFO members were mild reformers who merely wished to tinker with the system in the hope of obtaining a "proper" balance. All these elements certainly found expression in the movement, but there was something more, much more: UFO members, if only momentarily, glimpsed an alternative way of conducting human affairs and they strove to implement their vision. If radicalism was not completely realized in the movement, then at least the *potential* for radicalism existed.

UFO members saw themselves as an exploited group. They cast a critical eye upon the society in which they lived and were not only frustrated and disgusted with much of what they saw but also dissatisfied with how they were perceived and treated. They witnessed legislative bodies composed of lawyers and businessmen who had no understanding of the concerns of common people making laws on their behalf. Indeed, they believed that these legislators had nothing but contempt for the rank and file; hence the democracy that was so often boasted about was, in their view, a myth. They saw a press that routinely lied in the interests of economic and political power and often distorted the UFO's platform. They witnessed an urban culture that treated them as if they were semiliterate rubes. They observed an economic system in which they had little, if any, agency in determining the price they received for their labour. Urban-based manufacturers, propped up with tariff protection, gouged farmers for their implements and for the necessities of life while also stealing from the public treasury. Finally, they saw a society that did not consider more than half of its population – women – to be persons in the legal (or even the figurative) sense of the word. These non-persons were a vital part of farm operations, yet they had to struggle merely to obtain the vote.

In seeing all of this, UFO members in effect announced that they discerned the reality of the situation. In a society that incessantly boasted about the superiority of its democratic structures, they took a contrary position, arguing that the existing structures were not democratic, or at least not as democratic as they should be. UFO members were willing to take action to effect change. Farmers did not turn their backs on the problem and let "great men" lead them to "liberation," nor did they trumpet the capacity of the free market to set things right. They spontaneously determined to act for themselves. Thus, UFO members fielded candidates who were pledged to legislate in the interests of "the People," an abstraction used to describe the mass of humanity that did not enjoy political or economic power to any meaningful extent. In addition, they formed new cooperatives and built upon existing ones, not only to maximize returns but also to demonstrate that there was an alternative way to conduct commercial affairs. In sum, they saw things differently from mainstream society and, more importantly, attempted to build institutions designed to remedy many of the wrongs they perceived.

The question whether rank-and-file UFO members saw the movement as all-inclusive or purely as an agrarian force is difficult to answer definitively. From the evidence presented here, it appears that the issue was never successfully resolved at the local level. In fact, aside from the occasional reference to the Drury/Morrison split, it was not seen as an important concern by the rank and file. At times, UFO members jealously guarded their organization from encroachment by outside groups, such as labour; on other occasions, they made sincere and exhaustive efforts to align themselves with urban forces in order to contest elections. Whether or not they should formally join with these groups was a strategic concern, however, rather than a matter of principle. As already noted, many UFO members conceptualized society as consisting of two broad groups – the Big Interests and the "People." Their goal was to wrest power away from the former group so that the quality of life for the latter would be more meaningful and enjoyable. Although this might seem to necessitate formal or informal alliances among groups that perceived themselves to be oppressed, agrarians for the most part did not conceive of the formation of such relationships as a critical issue. The primary interest of UFO members was to address the concerns of farmers.

An inequitable system functions without blatant coercion only when the people who do not share in societal power still believe that the system is fair. If individuals become aware of the contradictions that surround them and form organizations to challenge inequities, then efforts must be made to bring these people back into the fold. Attempts to bring the UFO back into the mainstream were not overt, and certainly there was no conscious conspiracy in this regard. Even so, such efforts existed and most often took the form of promoting and reinforcing the key assumptions that underpinned the *status quo*.

When they elected MPPs and MPs to Toronto and Ottawa, farmers discovered exactly how powerful the old order was. UFO politicians soon found themselves arguing for calm, pragmatic solutions to problems rather than for fundamental change. Even the alliance with the most natural ally, the Independent Labor Party, broke down. Compounding the situation was the rather ambiguous position cooperatives came to occupy, thanks in part to the businesslike attitude of some of the movement's leaders and in part to state intervention. It is small wonder, then, that the movement lost much of its drive and enthusiasm in such a relatively short period of time. Seeing that it did not matter much that they had built a political force capable of capturing power at Queen's Park and exerting pressure at the federal level, many farmers became disillusioned and gave up. The expectations of UFO members were simply not met.

This study has focused on the democratic tendencies evident in the behaviour of UFO members. In current usage, democracy has lost much of its meaning. It is chiefly employed as a catchword by those in positions of power who wish to score political points. A related word, "populist," has been all but appropriated by the political right and is now used to describe leaders such as Ronald Reagan, Preston Manning, Mike Harris, and a host of others who do not talk so much about common people as they do about the need to unfetter the private sector. UFO members, however, interpreted democracy differently; they saw it as a mechanism that allowed all people to participate in the collective decisions that shape a society.

Implicit in this interpretation of the UFO and its members are many of the attributes found in anarchist theory. As L. Susan Brown notes, anarchists are those who "affirm a commitment to the primacy of individual freedom" in conjunction with the belief that "human individuals ... are best suited to decide for themselves how to run the affairs of their own lives; they are best served when left unconstrained by authority and unhampered by relationships of domination. The ontological basis for these beliefs is an understanding that individuals are free and responsible agents who are fit to determine their own development."[8] In the preceding chapters we have witnessed these traits – often unconscious – in many who comprised the UFO. We have seen a faith in the inherent value of all human beings; a belief that all people are, by nature, creative and competent enough to participate in decision-making processes; a suspicion of power; a commitment to cooperation; a realization that collective action is not possible until individuals break free from coercive power structures and join together to eliminate these structures; a conviction that democratic forms work most effectively when broken down into small units, which may then federate through free association. This is not to suggest that UFO members marched into meeting halls with the collected works of Bakunin under their arms. It is to suggest, however, that in the UFO one sees the historical tendency towards freedom that anarchists still expound.

Some scholars will persist in dismissing the UFO as an organization composed of cranky farmers who harboured an irrational fear of progress and paranoid delusions about a shadowy force they referred to as the Big Interests. But how inaccurate was the analysis put forward by UFO members? To what extent can their social critique be applied to present-day circumstances? How paranoid was it for UFO members to argue that those with capital effectively dictated public policy? Is there no resemblance between recent events and the period when the Smiths Falls Progressive Committee caustically observed that old-party politicians prate about a pay cheque and a full dinner pail, as if that was all human beings needed? Was it all that ridiculous for UFO members to allege that the old parties scapegoated certain groups as being the cause of economic and social problems in order to score political points? How inaccurate was the UFO complaint that the corporate interests had the old-party politicians in their back pockets?

Raising questions and critically assessing one's society is an arduous task. The exercise is made all the more difficult because human beings are exposed to countless messages (mostly reinforcing the *status quo*) on a daily basis. This difficultly notwithstanding, UFO members developed a probing critique of their society and formulated various means to attain a better quality of life for everyone. To be sure, the vision they devised contained mixed and contradictory elements, but to demand consistency in such a large project is unrealistic.

The central thrust of this book has been to demonstrate the extent to which rank-and-file UFO members were able to posit an alternative vision. Lambton, Simcoe, and Lanark UFO members struggled to advance a model of society that was quite different from what existed. Although the ability to posit an alternative vision evaporated in fairly short order, the UFO stands as an example of what common people can accomplish when the effort is made.

If one concedes that there is continuity between the past and present, then the experience of the UFO has relevance beyond its lifespan. The movement cannot serve as an exact blueprint for contemporary and future action. Yet if people can draw upon its experience and build upon its strategies for contending with inequitable power structures, then the efforts of UFO members were not without meaning beyond their generation. At the very least, their accomplishments should be perpetuated in the minds of Ontarians as a reminder that, in the event that change is desired, there are ways to accomplish it other than to give the free market unrestricted reign, or to adopt Marxist principles, or to grant political leaders even greater power. By referring to the experience of the UFO, populists may learn what traps to avoid, and they may see that the values they esteem were once held by people equally committed to transforming Ontario into a more democratic society.

In the early part of the twentieth century, there were many people who were not content with the prevailing state of affairs, and who tried to effect change. As UFO members discovered, the chances of success were slim, but they persisted nonetheless. Whether or not a populist movement with similar values will ever again emerge in Ontario is an unknown. One thing, however, can be stated with relative certainty: a mass democratic movement with the creativity, humanity, and democratic spirit of the United Farmers of Ontario will arise only when people muster the same capacity and courage shown by the UFO rank and file to see things differently.

Appendices

APPENDIX A

UFO Members Identified in Plympton Township, Lambton County

Name	*Age*	*Acres*	*Cleared*	*Nonprod*[1]	*$Land*	*$Buildings*	*$Total*
Dawson, Alex	61	50	40	10	1,450	600	2,050
Stonehouse, Joseph	47	70	60	10	1,900	400	2,300
Stonehouse, Roy	32	100	100		2,900	700	3,600
Smale, Silas C.	54	100	80	20	2,900	900	3,800
Smith, Charles	38	100	80	20	2,900	700	3,600
Jackson, James	45	100	80	20	2,900	800	3,700
Capes, Jonathan[2]	23	100	80	20	2,700	750	3,450
Park, Robert	65	97	97		2,700	500	3,200
Simpson, Thomas	51	98	83	15	2,850	800	3,650
Dewar, James		98	83	15	2,850		2,850
McPhedran, Roy[3]	37	98	83	15	2,850	900	3,750
McNeil, John C.	63	98	83	15	2,200	400	2,600
Ellwood, Joseph	48	199	174	25	5,600	800	6,400
Brownlee, David[4]	52	150	120	30	4,100	800	4,900
Thompson, Albert[5]	35	75	65	10	2,150	500	2,650
Ramsay, Orville[6]	33	100	80	20	2,900	900	3,800
McPhedran, Peter[7]	49	100	95	5	2,900	550	3,450
Forbes, William		50	40	10	1,450	300	1,750
Kerrigan, Dennis	38	50	50		1,400	500	1,900
McLean, James W.	50	100	95	5	2,900	750	3,650
Sanders, Fred	53	60	50	10	1,700	400	2,100
Williamson, Arch.	49	50	50		1,450	600	2,050
Bryson, Joseph[8]	58	150	120	30	4,100	400	4,500
Dewar, George A.	39	100	80	20	2,900	900	3,800
Hodgins, John	47	100	80	20	2,900	500	3,400
Hodgins, Calvin[9]	52	200	180	20	5,800	800	6,600
Hillier, John	61	175	155	20	5,050	850	5,900

Name	*Age*	*Acres*	*Cleared*	*Nonprod*[1]	*$Land*	*$Buildings*	*$Total*
Hillier, William	28	100	90	10	2,900	700	3,600
McLean, John G.		100	85	15	2,900	750	3,650
Ferguson, Duncan	36	150	125	25	4,300	1,250	5,550
Shea, Robert J.	31	50	40	10	1,450	500	1,950
Watson, Gordon[10]	24	150	120	30	4,350	800	5,150
Watson, David	66	50	40	10	1,450	600	1,850
Pascoe, Roy	34	100	95	5	2,900	650	3,550
Smith, William B.[11]	31	100	90	10	2,900	600	3,550
Scoffin, William[12]	46	106	96	10	3,050	650	3,650
Dewar, Arch.	43	50	40	10	1,450	500	1,950
Burnley, Thomas[13]	58	173	138	35	4,900	800	5,700
Jardine, Wm	41	150	130	20	4,350	800	5,150
Simpson, Kate[14]	27	104	84	20	3,000	700	3,700
Hoskins, Marshall[15]	30	147	137	10	3,900	400	4,300
Abell, Joseph[16]	59	150	130	20	4,300	700	5,000
Smith, Frend[17]	32	70	50	20	1,700		1,700
Galbraith, Dan W.[18]	31	150	135	15	4,000	850	4,850
Vanderburg, Earl[19]	21	100	90	10	2,600	600	3,200
McMillan, Arch.	55	90	80	10	2,600	800	3,400
Greenless, David	45	100	90	10	2,900	600	3,500
Sparling, John[20]	23	100	90	10	2,900	800	3,700
Wilkinson, Alex[21]	43	300	170	30	8,700	700	9,500
Tremaine, John[22]	47	150	120	30	4,150	700	4,850
Smith, Jonathan		100	100		2,900		2,900
Douglas, Herbert	47	200	180	20	5,800	500	6,300
McLaren, Angus	55	100	90	10	2,900	750	3,650
McLaren, Robert	57	97	97		2,800	500	3,300
Stogdill, Albin	36	100	90	10	2,900	700	3,600
Smith, Duncan[23]	54	399	319	80	11,100	1,200	12,300
Cairns, William H.	38	145	125	20	4,000	500	4,500
Scott, William		100	90	10	2,850	250	3,100
Fawcett, Henry		50	50		1,350		1,350
Campbell, Fred	36	80	70	10	2,300	700	3,000
Skinner, Richard[24]	49	70	70		2,450		2,450
Elliott, Huron	31	100	90	10	2,800	400	3,200

AGE AND HOLDINGS, PLYMPTON TOWNSHIP

UFO Members

Total No. UFO members identified	62
Average age	43.46
Average acres owned	115.41
Average acres cleared	98.46
Average acres waste	15.25
Average value, land	$3,273.73
Average value, buildings	$633.05
Average total value	$3,906.78

General	
Total of sample surveyed	231
Average age	48.29
Average acres owned	103.42
Average acres cleared	89.56
Average acres waste	13.85
Average value, land	$2,909.31
Average value, buildings	$603.07
Average total value	$3,512.38

Source: Lambton County Archives, Assessment Rolls – Plympton Township, 1922

Notes:

1 For appendices A to P, this category includes woodland, slash, and swamp; and other wasteland.

2 Listed under James Capes, his father, who was sixty-five and at that time considered to be in control of the land.

3 Also listed as owner is Mary R. McPhedran, age twenty-nine, and apparently Roy's sister, given their respective ages.

4 Includes fifty acres (value $1,200) that belonged to his wife, Susie.

5 Includes seven hundred acres (value $700) that belonged to his wife, Lena.

6 Also listed is his wife, Mary, also a UFO member.

7 Also listed is his wife, Catharine, also a UFO member.

8 Includes fifty acres (value $1,200) that belonged to his wife, Isabelle.

9 Includes one hundred acres (value $2,900) that belonged to his wife, Agnes.

10 Property is listed under William Watson, his father. Gordon also had two brothers listed on the assessment roll (ages thirty-one and twenty-two). If all three stood to inherit some of the land, then Gordon's share may have been only a percentage of the total holdings. Figures also include fifty acres (value $1,450) that belonged to William Watson's wife, Emily.

11 Smith was a tenant.

12 Includes six acres (value $150) that belonged to his wife, Gertrude.

13 Includes seventy-five acres (value $2,100) that belonged to his wife, Margaret.

14 Kate Simpson was listed as owning four acres (value $100) and was married to Jonathan Simpson. Figures represent their combined holdings.

15 Hoskins lived on forty-seven acres (value $1,200) that adjoined his widowed parents' holdings. Figures represent the combined holdings.

16 Includes fifty acres (value $1,400) that belonged to his wife, Dinah.

17 Smith was a tenant. Some newspaper accounts spell his first name "Friend."

18 No property was listed as belonging to Dan W. Galbraith. He and his wife lived with his father, Jonathan. Figures represent the father's holdings.

19 Earl Vanderburg had no property listed as his own. Figures represent the total holdings of his parents, Joseph and Ida.

20 Although John Sparling is listed, it is not as owner. He lived with his widowed mother, Ellen, and the figures represent her holdings.

21 Includes two hundred acres (value $5,800) that belonged to his wife, Harriett.

22 Includes fifty acres (value $1,250) that belonged to his wife, Eunice.

23 Includes 199 acres (value $5,600) that belonged to his wife, Lillie.

24 Skinner was a tenant.

APPENDIX B

UFO Members Identified in Bosanquet Township, Lambton County

Name	*Age*	*Acres*	*Cleared*	*Nonprod*	*$Land*	*$Buildings*	*$Total*
Richardson, Cliff	28	68	64	4	2,300	700	3,000
Stonehouse, George	62	86	80	6	3,250	850	4,100
Isaac, Ardeau	56	59	59		2,225	750	3,225
Smith, George	38	100	95	5	3,400	1,100	4,500
Wight, Roland	25	100	85	15	3,400	850	4,250
Finnie, William[1]	71	62	62		1,700	150	1,850
Zavitz, James	53	50	45	5	1,400	450	1,850
Murray, Andrew	50	2.5	2.5		100	250	350
Moloy, Earl[2]	25	154	37	117	2,300	175	2,475
Wilson, John	30	60	60		2,050	1,150	3,200
Hamilton, John[3]	24	40	30	10	800		800
Marrison, Preston	49	112.5	82.5	30	2,900	775	3,675
Tetzell, Roy[4]	36	100	85	15	3,400	1,200	4,600
Powell, John H.	56	105	105		3,700	300	4,000
Jamieson, Archie	49	50	25	25	1,400		1,400
Moloy, Thomas	62	66.5	60.5	6	2,250		2,250
Moloy, Fred	28	50	50		1,500		1,500
Sitter, Nicholas[5]	47	292	237	55	10,150	2,000	12,150
Clark, John	46	10	10		350	25	375
McIntyre, John[6]	39	100	95	5	3,400	1,100	4,500
Carmichael, Adam	43	150	146	4	5,100	1,150	6,250
Wilsie, Jacob	63	113	113		3,500	1,400	4,900
Carrothers, Isaac	63	201	186	15	4,875	950	5,825
Shepherd, Fred	44	150	150		5,700	900	6,600
Elliott, Moses	59	50	50		1,900	975	2,875
Gilliard, James	48	100	95	5	3,400	1,050	4,450

Name	Age	Acres	Cleared	Nonprod	$Land	$Buildings	$Total
French, William	40	150	150		5,400	1,350	6,750
Lawrie, James A.	51	99	90	9	3,150	1,000	4,150
Wells, Burton	44	72	47	5	2,675	1,000	3,675
Lithgow, Robert[7]	32	100	90	10	3,400	1,350	4,750
Stewart, David[8]	44	100	100		3,400	50	3,450
Tidball, Robert	52	100	100		3,400	1,350	4,750
Hare, Albert	63	141	131	10	4,640	1,350	5,990
Frayne, Isaac	60	100	100		3,600	1,000	4,600
Frayne, Earl	25	100	100		3,700	800	4,500
Valentine, Ebenezer	29	200	190	10	7,000	850	7,850
Rawlings, Fred	52	89.5	89.5		3,800	900	4,700
Stutt, Richard	54	200	185	15	6,800	1,150	7,950
Dew, John	51	200	192	8	7,200	1,050	8,250
Wellington, Garner	36	100	100		3,800	1,000	4,800
Dew, Frederick	37	100	95	5	3,800	1,050	4,850
McIntyre, Duncan	62	50	50		1,900	900	2,800
Blundon, Sanford	59	50	50		1,900		1,900
Whyte, Duncan	54	100	80	20	3,800	1,200	5,000
Dew, Frank	33	108	80	28	4,100	950	5,050
Taylor, David	54	125	80	45	3,550	1,300	4,900
Vivian, John	69	117.5	117.5		4,450	1,000	5,450
Vance, Arthur	45	133	100	33	4,700	750	5,450
Lester, George	69	121	80	41	3,500	1,100	4,600

AGE AND HOLDINGS, BOSANQUET TOWNSHIP

UFO Members	
Total no. UFO members identified	49
Average age	47.12
Average acres owned	105.26[9]
Average acres cleared	95.40
Average acres waste	9.41
Average value, land	$3,602.61
Average value, buildings	$865.34
Average total value	$4,467.95
General	
Total of sample surveyed	168
Average age	47.06
Average acres owned	108.34
Average acres cleared	93.80
Average acres waste	14.53
Average value, land	$3,505.36
Average value, buildings	$906.69
Average total value	$4,412.05

Source: Lambton County Archives, Assessment Rolls – Bosanquet Township, 1922

Notes:

1 Finnie is listed as a tenant.

2 Moloy is listed as a tenant.

3 Hamilton is listed as a tenant.

4 Tetzell is listed as a tenant.

5 Includes 152 acres listed as owned by his wife Ellen (value $5,200).
6 McIntyre is listed as a tenant.
7 Lithgow apparently co-owned the property with his father, Thomas.
8 Stewart's property adjoined that of his widowed mother, Margaret. The total value of his mother's holdings was $4,700.
9 Figures related to tenant farmers were not used in determining this average.

APPENDIX C

UFO Members Identified in Enniskillen Township, Lambton County

Name	*Age*	*Acres*	*Cleared*	*Nonprod*	*$Land*	*$Buildings*	*$Total*
Munro, Jonathan	60	100	80	20	2,800	850	3,650
Doolan, William		200	100	100	4,100		4,100
Simpson, Thomas		75	60	15	2,250		2,250
Wilkinson, George	40	131	111	20	2,800	600	3,400
Hall, Franklin	27	100	40	60	2,300	300	2,600
Shortt, George	44	200	180	20	3,500	600	4,100
Thompson, Harvey	35	150	140	10	4,300	800	5,100
Annett, George	49	150	130	20	3,800	800	4,600
Napper, Russell		42.5	30	12.5	1,150		1,150
Anderson, Ernest	32	100	80	20	2,400	600	3,000
Mackesy, Jonathan		50	20	30	900		900
Simpson, Thomas	70	100	50	50	2,700		2,700
Wilson, Wilfred	23	98	60	38	2,400	300	2,700
Leith, William		67	50	17	1,800	500	2,300
Sharp, William D.	56	100	30	70	2,400	200	2,600
Farrow, Isaac		61	40	40	1,150	300	1,450
Lecocq, William J.	57	200	150	50	6,000	1,200	7,200
McLennan, James		100	90	10	1,700		1,700
Park, Robert J.[1]		100	80	20	3,000	800	3,800
Welch, James E.[2]	70	200	160	40	5,800	800	6,600
Brock, W. Albert	54	100	80	20	2,000	600	2,600
Watt, William	52	75	70	5	2,300	600	2,900
Hackett, Joseph	47	100	90	10	3,200	1,000	4,200
Williams, James	51	100	90	10	2,900	900	3,800
Brock, Thomas		100	90	10	3,100	400	3,500
Wright, Daniel	51	100	90	10	3,100	800	3,900
Currah, Joesph	48	100	80	20	2,900	600	3,500

Name	*Age*	*Acres*	*Cleared*	*Nonprod*	*$Land*	*$Buildings*	*$Total*
Currah, James	50	100	70	30	2,800	800	3,600
Smith, Charles		100	80	20	2,200		2,200
Stonehouse, Angus	44	100	90	10	3,100	700	3,800
Anderson, Harvey	30	100	90	10	3,100	800	3,900

AGE AND HOLDINGS, ENNISKILLEN TOWNSHIP

UFO Members	
Total no. UFO members identified	31
Average age	47.77
Average acres owned	108.98
Average acres cleared	83.90
Average acres waste	25.08
Average value, land	$2,837.10
Average value, buildings	$511.29
Average total value	$3,348.39
General	
Total of sample surveyed	108
Average age	48.13
Average acres owned	101.98
Average acres cleared	78.77
Average acres waste	23.21
Average value, land	$2,762.04
Average value, buildings	$594.44
Average total value	$3,356.48

Source: Lambton County Archives, Assessment Rolls – Enniskillen Township, 1922

Notes:

1 Listed as co-owner with Arthur Park (relationship unknown).

2 Listed as co-owner with Harold Brock.

APPENDIX D

UFO Members Identified in Warwick Township, Lambton County

Name	*Age*	*Acres*	*Cleared*	*Nonprod*	*$Land*	*$Buildings*	*$Total*
McPherson, Vaughn[1]	25	50	50		2,000		2,000
Cates, Jacob	56	140	135	5	5,900	800	6,700
Curts, Gordon	22	100	95	5	4,200	800	5,000
Farrell, John	60	150	135	15	6,200	400	6,600
Kernohan, Basil[2]	21	100	95	5	4,000		4,000
Ellerker, Fred	27	50	50		2,100	500	2,600
Karr, Ernest	44	100	85	15	4,200	600	4,800
Brandon, James	44	100	90	10	4,200	700	4,900
Brandon, William	42	100	90	10	4,200	400	4,600
Brandon, Robert	39	100	85	15	4,200	500	4,700
Brent, George	49	150	140	10	6,300	700	7,000
Brent, Stanley	39	200	190	10	8,400	700	9,100
Lester, George	42	100	100		4,200	300	4,500
Tomlinson, Arthur	45	50	50		2,100	200	2,300
Luckham, W.H.	64	130	120	10	5,950	800	6,750
Vance, William J.	45	150	140	10	6,200	800	7,000
Vance, Gordon[3]	27	2	2		150	500	650
Hall, Robert A.	38	150	140	10	6,200	1,050	7,250
Luckham, Macklin	55	100	90	10	4,200	700	4,900
Scoffin, John	50	100	85	15	4,000	500	4,500
Graham, Fred	24	100	80	20	4,200	800	5,000
Thompson, Allan	37	100	90	10	4,000	700	4,700
Hall, Lloyd	23	100	100		4,100		4,100
Auld, Andrew	50	305	265	40	11,800	950	12,750
Janes, C.E.	32	150	130	20	6,300		6,300
Yorke, Basil	41	150	130	20	6,300	800	7,100
Wilkinson, John C.	50	250	205	45	10,500	1,600	12,100
Young, George	34	75	75		3,000	600	3,600

AGE AND HOLDINGS, WARWICK TOWNSHIP

UFO Members	
Total no. UFO members identified	28
Average age	40.18
Average acres owned	124.07
Average acres cleared	112.60
Average acres waste	11.48
Average value, land	$5,146.30
Average value, buildings	$588.89
Average total value	$5,735.19
General	
Total of sample surveyed	120
Average age	48.29
Average acres owned	117.95
Average acres cleared	107.46
Average acres waste	10.49
Average value, land	$4,779.17
Average value, buildings	$644.17
Average total value	$5,423.33

Source: Lambton County Archives, Assessment Rolls – Warwick Township, 1922

Notes:

1 Listed as living on land adjacent to the holdings of his brother, Charlie, who owned property and buildings valued at $9,400.

2 Listed as co-owner with brother, Stanley. Father, David, lived on adjoining lot, with property and buildings assessed at $7,000.

3 Listed as co-owner wih father, Ezekiel, whose occupation is listed as merchant. Property value is not included in total calculations.

APPENDIX E

UFO Members Identified in Brooke Township, Lambton County

Name	*Age*	*Acres*	*Cleared*	*Nonprod*	*$Land*	*$Buildings*	*$Total*
Fisher, Duncan	63	100	90	10	2,800	500	3,300
Campbell, Arch.[1]	24	100	90	10	2,900	700	3,600
Darvill, John	60	50	50		1,700	300	2,000
Oke, Leslie W.[2]	43	50	50		1,700	700	2,400
Gilroy, Henry A.	53	228	200	28	6,900	600	7,500
McIntyre, Arch.	46	100	80	20	2,500	500	3,000
Clark, John	55	50	40	10	1,600	400	2,000

AGE AND HOLDINGS, BROOKE TOWNSHIP

UFO Members

Total no. UFO members identified	7
Average age	49.14
Average acres owned	96.86
Average acres cleared	85.71
Average acres waste	11.14
Average value, land	$2,871.43
Average value, buildings	$528.57
Average total value	$3,400.00

Source: Lambton County Archives, Assessment Rolls – Brooke Township, 1921

Notes:

1 Listed as co-owner with father, George.

2 Served as UFO MPP, 1919 to 1929.

APPENDIX F

UFO Members Identified in Moore Township, Lambton County

Name	*Age*	*Acres*	*Cleared*	*Nonprod*	*$Land*	*$Buildings*	*$Total*
Young, Byron	28	100	80	20	3,800	1,000	4,800
Curran, Edward	57	100	90	10	3,900	1,500	5,400
White, Robert J.	41	75	60	15	3,000	1,100	4,100
Johnson, Andrew	52	100	80	20	4,000	1,100	5,100
McMahon, Fred J.[1]	63	100	90	10	4,200	1,400	5,600

AGE AND HOLDINGS, MOORE TOWNSHIP

UFO Members

Total no. UFO members identified	5
Average age	48.20
Average acres owned	95.00
Average acres cleared	80
Average acres waste	15
Average value, land	$3,870.00
Average value, buildings	$1,220.00
Average total value	$5,000.00

Source: Lambton County Archives, Assessment Rolls – Moore Township, 1922

Note:

1 Listed as co-owner with son, James F.

APPENDIX G

UFO Members Identified in Euphemia Township, Lambton County

Name	Age	Acres	Cleared	Nonprod	$Land	$Buildings	$Total
Annett, Harvey[1]	29	150	145	5	2,800	700	3,500
Bailey, John R.[2]	42	77	77		1,950	350	2,300
Fansher, Bert W.[3]	41	225	180	45	5,025	450	5,475

AGE AND HOLDINGS, EUPHEMIA TOWNSHIP

UFO Members

Total no. UFO members identified	3
Average age	37.33
Average acres owned	150.67
Average acres cleared	134.00
Average acres waste	16.67
Average value, land	$3,258.33
Average value, buildings	$500.00
Average total value	$3,758.33

Source: Lambton County Archives, Assessment Rolls – Euphemia Township, 1922

Notes:

1 Listed as co-owner with wife, Sarah.

2 Listed as co-owner with wife, Margaret.

3 Served as MP for East Lambton, 1921 to 1925; 1926 to 1930.

APPENDIX H

UFO Members Identified in Oro Township, Simcoe County Oro Station UFO Club

Name	*Age*	*Acres*	*Cleared*	*Nonprod*	*$Land*	*$Buildings*	*$Total*
McArthur, Alex	45	102	77	25	1,800	800	2,600
Ross, Hugh	43	100	90	10	1,800	700	2,500
Kirkpatrick, Geo.[1]	47	221	171	50	3,700	1,300	5,000
Ross, Alexander[2]	26	220	205	15	3,400	1,300	4,700
Ross, Victor	22						
Bell, Allan	38	100	90	10	1,900	700	2,600
Bell, Thomas[3]	32	111.5	96.5	15	2,100	1,500	3,600
Currie, Alex	31	100	90	10	1,800	800	2,600
Luck, Thomas	52	163	153	10	3,000	1,800	4,800
Fletcher, Alex J.	44	150	125	25	2,300	800	3,100
McArthur, William	63	50	20	30	500	300	800
Livingston, John[4]	38	100	80	20	700	400	1,100
Strachan, Geo.[5]	24	372	280	92	4,000	1,400	5,400
McArthur, John A.[6]	58	150	100	50	2,000	900	2,900
McArthur, Archibald	48						
Crawford, Albert[7]	28	125	115	10	2,900	1,200	3,100
Pearsall, Ben[8]		150	150		2,900	1,200	4,100
Crawford, George[9]	67	119.5	110.5	9	2,300	1,300	3,600
Crawford, Ernest	26						
Crawford, Fred[10]	38	162.5	107.5	55	2,550	1,000	3,550
Crawford, Wesley	45	162.5	107.5	55	2,550	1,000	3,550
McArthur, James[11]	21	100	80	20	1,300	600	1,900
Coates, Ernest[12]	23	150	100	50	1,900	800	2,700
Hickling, Ralph[13]	26	50	50		750	400	1,150
McArthur, William[14]		5	5		100	500	600
Reid, John	30	175	145	30	2,300	1,000	3,300
Walker, John W.	46	183	115	68	2,500	900	3,400

Name	Age	Acres	Cleared	Nonprod	$Land	$Buildings	$Total
Emms, Joseph[15]	24	176	150	26	2,600	400	3,000
Wiggins, John	54	100	90	10	1,800	1,000	2,800
Ross, Thomas E.[16]	45	200	140	60	3,300	1,300	4,600
Kissock, Samuel	58	150	140	10	2,600	1,000	3,600
McArthur, Archibald	39	70	70		1,200	600	1,800
Gilchrist, George	50	125	80	45	1,150	550	1,700
Miller, Wilson	48	117	97	20	2,100	900	3,000

AGE AND HOLDINGS, ORO STATION UFO CLUB, ORO TOWNSHIP

UFO Members	
Total no. UFO members identified	34
Average age	34.24
Average acres owned	139.53
Average acres cleared	83.33
Average acres waste	56.19
Average value, land	$2,299.72
Average value, buildings	$657.78
Average total value	$2,957.50
General	
Total of sample surveyed	114
Average age	49.90
Average acres owned	159.38
Average acres cleared	97.59
Average acres waste	59.42
Average value, land	$2,575.18
Average value, buildings	$794.82
Average total value	$3,370.00

Source: Simcoe County Archives, Assessment Rolls – Oro Township, 1922

Notes:

1 Listed as co-owner with father, Guy.

2 Alexander Ross and his brother, Victor (see below), are listed as co-owners with father, Thomas, who was evidently not a member of the UFO. For the purposes of this table, all property value has been ascribed to Alexander. Other than for his age, Victor does not figure into the calculations.

3 Listed as co-owner with father, Arthur.

4 Listed as co-owner with Flora Revie.

5 Listed as co-owner with father, Robert.

6 Listed as co-owner with Archibald A. McArthur, who apparently was his brother and also an Oro Station UFO member (see below).

7 Listed as co-owner with father, Robert.

8 Listed as co-owner with Arthur Bell.

9 Listed as co-owner with son, Ernest, also an Oro Station UFO Club member (see below).

10 Listed as co-owner with father, Richard, and brothers, Albert (age thirty-two), Charles (age twenty-seven) and Wesley (age forty-five). In this table, the total value of the property is divided between Fred and Wesley.

11 Listed as co-owner with father, John J.

12 Listed as co-owner with father, James.

13 Listed as co-owner with Mary McLarty, who is referred to only as a "spinster." Since Hickling is listed as a farmer and co-owner of the property, his statistics have been included in the total calculations.
14 Listed as a trustee in the assessment roll. Statistics associated with William McArthur are not included in the final calculations.
15 Listed as co-owner with father, William.
16 Although Ross was not a member of the Oro Station UFO, he served as Progressive MP for the riding of North Simcoe, 1921–25.

APPENDIX I

UFO Members Identified in Oro Township, Simcoe County Rugby UFO Club

Name	*Age*	*Acres*	*Cleared*	*Nonprod*	*$Land*	*$Buildings*	*$Total*
Anderson, David C.	31	225	190	35	3,300	1,200	4,500
Langman, Arthur[1]		136	136		2,100	1,200	3,300
Langman, Harry							
Ratcliff, Edward	50	150	140	10	2,100	900	3,000
Leigh, Jebez[2]	60	400	300	100	5,900	1,700	7,600
Leigh, Montgomery	25						
McLeod, Keith	25	100	80	20	1,000	500	1,500
Hoover, Wm, Jr[3]	42	148	148		2,200	1,000	3,200
Hoover, Wm, Sr.							
Scott, George	48	100	100		1,200	500	1,700
Locke, William J.	30	250	180	70	3,400	1,300	4,700
Horne, Benjamin[4]	50	200	180	20	3,100	1,000	4,100
Horne, James E.	53						
Buchanan, Donald	64	216	80	136	2,600	1,000	3,600
Johnstone, Elwood[5]	29	34	11	23	300	350	650
Horne, James H.[6]	42	248	198	50	3,800	1,300	5,100
Horne, William F.	38						
Fell, Eldred[7]	32	125	109	16	1,900	800	2,700
Jeremy, Thomas	40	166	126	40	2,450	900	3,350
Langman, Herbert	33	150	100	50	2,100	700	2,800
Johnstone, Edgar	36	98	70	28	1,400	700	2,100
Johnstone, Robert[8]	62	176	131	45	2,300	1,000	3,300
Johnstone, Wilfred	24						
Langman, George	45	100	100		1,800	1,200	3,000
Robertson, William	39	200	130	70	2,700	800	3,500
Anderson, Charles	36	125	75	50	1,225	875	2,100
Tudhope, Wesley		245	180	65	3,000	1,200	4,200

Name	Age	Acres	Cleared	Nonprod	$Land	$Buildings	$Total
Anderson, Andrew	25	215	195	20	2,200	700	2,900
Horne, Leonard	36	150	125	25	2,400	1,200	3,600
Moore, William	40	100	75	25	1,700	800	2,500

AGE AND HOLDINGS, RUGBY UFO CLUB, ORO TOWNSHIP

UFO Members	
Total no. UFO members identified	30
Average age	39.81
Average acres owned	174.91
Average acres cleared	136.87
Average acres waste	38.04
Average value, land	$2,429.35
Average value, buildings	$977.17
Average total value	$3,406.52
General	
Total of sample surveyed	114
Average age	49.90
Average acres owned	159.38
Average acres cleared	97.59
Average acres waste	59.42
Average value, land	$2,575.18
Average value, buildings	$794.82
Average total value	$3,370.00

Source: Simcoe County Archives, Assessment Rolls – Oro Township, 1922

Notes:

1 Listed as co-owner with son, Harry, also a Rugby UFO club member (see below).

2 Listed as co-owner with sons, Perry and Montgomery. Montgomery was also a member of the Rugby UFO club (see below).

3 Listed as co-owner with father William Sr, also a member of the Rugby UFO club (see below).

4 Listed as co-owner with James E. Horne, presumably his brother. James was also a member of the Rugby UFO club (see below).

5 Johnstone owned the property listed here, but "labourer" was used to denote his occupation. His property is not included in the members' averages.

6 Listed as co-owner with William F. Horne (age thirty-eight), presumably his brother. William was also a member of the Rugby UFO club (see below).

7 Listed as co-owner with father, James.

8 Listed as co-owner with son, Wilfred, also a Rugby UFO club member (see below).

APPENDIX J

UFO Members Identified in Flos Township, Simcoe County Edenvale UFO Club

Name	*Age*	*Acres*	*Cleared*	*Nonprod*	*$Land*	*$Buildings*	*$Total*
Ward, William	29	100	100		3,600	1,000	4,600
Giffen, Newman	35	185	185		9,400	1,200	10,600
Giffen, Henry	31	150	150		7,100	1,200	8,300
McDonald, Donald[1]	34	1	1		50	350	400
McNabb, Duncan	67	100	10	90	300	400	700
Culham, Wesley	52	100	90	10	4,100	1,300	5,400
Bowser, George	41	100	70	30	1,900	800	2,700
Rupert, James[2]	77						
Rupert, Zeeman	56	76	76		1,500	1,200	2,700
Rupert, Roy	28	78	70	8	1,900	500	2,400
Pearson, Horatus	59	70	40	30	1,700	200	1,900
Maw, William[3]	29	7	7		200	200	400
Maw, Albert	35	100	100		4,300	500	4,800
Maw, Wilfred	24	100	100		4,500	1,000	5,500
McNabb, John	72	135	50		85	1,500	1,500
Richardson, Abraham	47	176	85	91	1,800	1,000	2,800

AGE AND HOLDINGS, EDENVALE UFO CLUB, FLOS TOWNSHIP

UFO Members

Total no. UFO members identified	16
Average age	44.75
Average acres owned	113.08
Average acres cleared	86.62
Average acres waste	26.46
Average value, land	$3,353.85
Average value, buildings	$792.31
Average total value	$4,146.15

General	
Total of sample surveyed	102
Average age	44.61
Average acres owned	111.21
Average acres cleared	90.17
Average acres waste	21.04
Average value, land	$3,113.24
Average value, buildings	$806.86
Average total value	$3,920.10

Source: Simcoe County Archives, Assessment Rolls – Flos Township, 1922

Notes:

1 Since McDonald's occupation is listed as blacksmith, his figures are not included in the calculations.

2 Listed as retired, and as co-owner of property with son, Zeeman, also an Edenvale UFO club member (see below).

3 Since Maw's occupation is listed as bricklayer, his figures are not included in the calculations.

APPENDIX K

UFO Members Identified in Ramsay Township, Lanark County

Name	*Age*	*Acres*	*Cleared*	*Nonprod*	*$Land*	*$Buildings*	*$Total*
Dezell, William	60	250	180	70	2,900	1,000	4,000
Robertson, William[1]	58	163	72	91	1,725	1,100	2,825
Robertson, George	38	100	90	10	2,500	1,600	4,000
Dunlop, W.G.[2]	40	149.5	148.5	1	2,900	800	3,700
Burgess, George A.		100	80	20	1,000		1,000
Bowland, Byron		198	160	38	4,000	1,500	5,500
McCreary, Hiram		400	290	110	5,600	2,200	7,800
Turner, James[3]		100	75	25	1,000	500	1,500
Hilliard, Robert		100	80	20	1,500	600	2,100
Black, Robert[4]		200	135	65	1,200	600	1,800
Sutherland, Peter		260	228	32	4,000	1,200	5,200
Reid, J.R.		100	80	20	2,100	1,100	3,200
Sadler, T.J.		100	98	2	2,000	1,500	3,500
McArton, John		100	60	40	1,800	1,000	2,800
McArton, J.A.		100	100		2,000	800	2,800
Paul, W.J.		100	90	10	3,000	1,800	4,800
Kenny, James		100	85	15	2,500	1,500	4,000
Yuill, Robert[5]		125	101	24	2,700	1,200	3,900
Sutherland, Angus		100	75	25	1,000	1,000	2,000
Paterson, James[6]		250	125	125	3,400	600	4,000
Cochrane, Peter		100	85	15	3,500	1,300	4,800
Cochrane, Wilbert		100	80	20	2,000	2,000	
Chapman, Joe		144	120	24	2,525	1,300	3,825
Doherty, George[7]		100	86	14	2,200	800	3,000
Wylie, J.B.[8]		206 .5	105.5	101	3,750	3,400	7,150
Barker, Robert		100	88	12	3,550	1,000	4,550
Steele, John		100	70	30	4,750	1,500	6,250

Name	*Age*	*Acres*	*Cleared*	*Nonprod*	*$Land*	*$Buildings*	*$Total*
Cochrane, Andrew[9]		100	96	4	5,000	1,600	6,600
Gardner, William		132	80	52	2,000	1,000	3,000
Young, Mrs C.W.[10]		200	180	20	3,200	1,300	4,500
Matthews, Alton		100	90	10	3,500	700	4,200
Forde, Andrew		100	80	20	4,000	1,000	5,000
Snedden, W.A.[11]		158	98	60	3,300	1,500	4,800
Barker, Alex		117	117		3,900	600	4,500
Robertson, W.H.		75	75		1,700	600	2,300
Wright, James		100	60	40	3,300		3,300
McGill, John[12]		80	70	10	2,000	500	2,500
Ryan, John		100	100		5,000	2,000	7,000
McPhail, Dan		100	80	20	1,400	100	1,500
Syme, Peter		190	140	50	4,000	1,500	5,500
Stewart, Donald		175	125	50	5,225	800	6,025
Neilson, W.H.		200	120	80	5,500	1,500	7,000
Ryan, Michael		600	347	253	2,900	1,200	4,100
Drummond, Sam		100	71	29	3,500	1,500	5,000
Ross, Fred A.		100	85	15	3,000	1,000	4,000
Curtin, Lawrence[13]		100	60	40	400		400
Jamieson, R.		100	70	30	500	200	700

AGE AND HOLDINGS, RAMSAY TOWNSHIP

UFO Members	
Total no. UFO members identified	47
Average aces owned	147.24
Average acres cleared	109.67
Average acres waste	37.56
Average value, land	$2,874.46
Average value, buildings	$1,058.87
Average total value	$3,933.15
General	
Total of sample surveyed	114
Average acres owned	158.53
Average acres cleared	112.91
Average acres waste	44.73
Average value, land	$2,787.68
Average value, buildings	$876.54
Average total value	$3,664.21

Source: Archives of Ontario, Assessment Rolls – Ramsay Township, 1914

Notes:

1 Listed as sharing the property with son, Roy.

2 Listed as co-owner of property with father, John.

3 John and John S. Turner owned adjoining land with a total value of $4,000.

4 Listed as co-owner with Darrill Black. Because neither of these individuals' ages are listed, relationship is unknown. Even so, Darrill is listed as owning land and buildings worth $1,600, while Robert's share is only $200.

5 Twenty-five acres of land listed as belonging to R.M. Yuill, presumably his son, are also included in the assessment figures.

6 Listed as Patterson in most newspaper accounts.
7 Listed under property belonging to the estate of A. McIntyre. For the purposes of this table, Doherty is considered a tenant. Property figures are not included in total calculations.
8 Listed as a cheese manufacturer in the assessment roll, which explains the high value of buildings on his property.
9 Listed with Andrew Cochrane are Alex, Wilson, Robert, and Milton Cochrane. Alex and Milton were also members of the UFO.
10 Young was a widow.
11 Listed as co-owner with David Snedden, presumably his father. According to the assessment roll, W.A. Snedden owned sixty acres of land (none of which was cleared) worth $300. David Snedden owned ninety-eight acres (value $3,000) with buildings worth $1,500.
12 Listed as co-owner with Alex, Victor, and William McGill.
13 Listed as Curtain in most newspaper accounts.

APPENDIX L

UFO Members Identified in Drummond Township, Lanark County

Name	*Age*	*Acres*	*Cleared*	*Nonprod*	*$Land*	*$Buildings*	*$Total*
Cullen, Cecil[1]	27	100	60	40	2,100	900	3,000
McPhail, Donald[2]	21	250	115	135	3,400	700	4,100
Tetlock, Howard[3]	21	225	65	160	2,000	700	2,700
Shaw, Bland	29	100	80	20	2,200	500	2,700
Russell, W.J.[4]	21	100	50	50	1,800	650	2,450
Lewis, Ernest	21	100	50	50	1,600	500	2,100
Pennett, James[5]	24	100	80	20	2,400	700	3,100
Walsh, Patrick[6]	30	150	100	50	3,070	770	3,840
Dowdall, Richard	55	100	75	25	1,800	600	2,400
Dowdall, Lorne	28	100	50	50	1,600	190	1,790
Peters, George		100	60	40	1,400	350	1,750
Devlin, William	27	100	90	10	2,000	600	2,600
McTavish, Robert	44	136.5	100	36.5	3,950	900	4,850
McGregor, Peter	46	150	105	45	4,000	700	4,700
Ebbs, James	43	100	50	50	1,280	800	2,080
Somerville, James	63	100	80	20	1,700	650	2,350
Armstrong, Thomas[7]	25	200	100	100	1,500	500	2,000
Poole, Jonathan	57	300	190	110	3,595	1,130	4,725

AGE AND HOLDINGS, DRUMMOND TOWNSHIP

UFO Members	
Total no. UFO members identified	18
Average age	34.24
Average acres owned	139.53
Average acres cleared	83.33
Average acres waste	56.19
Average value, land	$2,299.72
Average value, buildings	$657.78
Average total value	$2,957.50
General	
Total of sample surveyed	114
Average age	49.90
Average acres owned	159.38
Average acres cleared	97.59
Average acres waste	59.42
Average value, land	$2,575.18
Average value, buildings	$794.82
Average total value	$3,370.00

Source: Archives of Ontario, Assessment Rolls – Drummond Township, 1920

Notes:

1 Listed as co-owner with father Henry and brother Meryl. Figures represent total holdings.
2 Includes 200 acres owned by father, Peter (value $3,200).
3 Includes 175 acres owned by father, Gilbert (value $2,540).
4 Listed as co-owner with father, Thomas.
5 Listed as co-owner with father, Louis.
6 Includes 100 acres owned by father, Daniel.
7 Listed as co-owner with father, James.

APPENDIX M

UFO Members Identified in Montague Township, Lanark County

Name	*Age*	*Acres*	*Cleared*	*Nonprod*	*$Land*	*$Buildings*	*$Total*
Bunting, Elias		94.5	94.5		2,600	600	2,600
Chalmers, Harry[1]		75	45	30	1,380	400	1,780
Condie, George	60	110	80	30	1,400	1,300	2,700
Condie, Robert	63	60	60		1,200	1,000	2,200
Condie, H.A.[2]	24	199	199		3,200	700	3,900
Shaw, Clarence[3]	23	50	25	25	600		600
Condie, Daniel	50	54.5	54.5		810		810
Carroll, Luke	77	67.5	50	17.5	900	200	1,100
Condie, E.R.	56	50	50		800	400	1,200
Code, William[4]	71	100	75	25	1,700	700	2,400
Edmunds, James E.[5]	58	81	81		2,200	800	3,000
Ferguson, J.D.	56	100	50	50	400		400
Farrell, James	62	100	90	10	1,400	400	1,800
McLenaghan, A.C.	53	127.5	125	2.5	4,300	1,100	5,400
McKenna, Chris	51	128	118	10	1,700	300	2,000
McPherson, Norman	35	100	80	20	600	200	800

AGE AND HOLDINGS, MONTAGUE TOWNSHIP

UFO Members

Total no. UFO members identified	16
Average age	52.79
Average acres owned	96.47
Average acres cleared	83.47
Average acres waste	13.00
Average value, land	$1,639.33
Average value, buildings	$540.00
Average total value	$2,179.33

General	
Total of sample surveyed	67
Average age	47.07
Average acres owned	188.46
Average acres cleared	118.06
Average acres waste	70.40
Average value, land	$2,010.22
Average vaule: buildings	$604.48
Average total value	$2,614.70

Source: Archives of Ontario, Assessment Rolls – Montague Township, 1920

Notes:

1 Chalmers is listed with what appears to be his father, Edmond. Since Harry's holdings are of such little value ($80), the total family holdings are included here.

2 Listed as holding adjoining land with George Condie, Jr (age twenty-nine), presumably his brother. The figures here are for the total holdings, given the relatively small value of the holdings of H.A.

3 Listed as a tenant. Property holdings are not included in total calculations.

4 Listed as co-owner of property with Thomas and John Code.

5 Includes ten acres owned by his wife, a UFWO member.

APPENDIX N

UFO Members Identified in Beckwith Township, Lanark County

Name	*Age*	*Acres*	*Cleared*	*Nonprod*	*$Land*	*$Buildings*	*$Total*
Henderson, Daniel	68	100	80	20	1,175	650	1,525
McEwen, R.J.[1]	37	150	110	40	1,800	650	2,450
McDiarmid, John	57	300	140	160	2,175	450	2,625
Scott, Robert A.	49	400	125	275	2,600	750	3,350
McTavish, Alex	52	200	80	120	1,700	750	2,450
Kettles, R.W.	31	100	80	20	1,400	350	1,750
Ferguson, Aberdeen	24	200	100	100	1,800	600	2,400
Ferguson, Mel.[2]	27	150	110	40	1,900	700	2,600
Carmichael, Norm[3]	38	180	80	100	1,700	500	2,200
McCuan, Thomas	47	175	155	20	2,750	1,000	3,750
McNeely, W.E.[4]	39	123	98	25	1,400	525	1,925
Timmons, Patrick	60	98	90	8	1,700	900	2,600
Simpson, Homer K.	42	136	98	38	1,450	700	2,150
Burgess, G.A.[5]		20	20		125		125
Cram, Willard	41	99	90	9	1,500	700	2,200

AGE AND HOLDINGS, BECKWITH TOWNSHIP

UFO Members

Total no. UFO members identified	15
Average age	43.71
Average acres owned	172.21
Average acres cleared	102.57
Average acres waste	69.64
Average value, land	$1,789.29
Average value, buildings	$658.93
Average total value	$2,448.21

General	
Total of sample surveyed	72
Average age	50.52
Average acres owned	213.13
Average acres cleared	128.92
Average acres waste	84.21
Average value, land	$1,635.07
Average value, buildings	$496.53
Average total value	$2,131.60

Source: Archives of Ontario, Assessment Rolls – Beckwith Township, 1917
Notes:
1 Listed as co-owner of property with father, Dan.
2 Figures for Mel. Ferguson include property owned by father, John. Melville's share of the property was fifty acres (value $600).
3 Figures include eighty acres owned by his widowed mother, Janet (Value $300).
4 Co-owned property with father, William.
5 Burgess was mayor of Carleton Place. Although he owned twenty acres of land (value $125), his total assessment came to $5,950, suggesting that he owned property elsewhere. The figures for Burgess are not included in the UFO township averages.

APPENDIX O

UFO Members Identified in North Elmsley Township, Lanark County

Name	*Age*	*Acres*	*Cleared*	*Nonprod*	*$Land*	*$Buildings*	*$Total*
Bowen, Robert[1]		150	140	10	2,200	500	2,700
Cullen, Mowat[2]		200	130	70	5,000	1,200	6,200
Frizzell, J.H.		50	50		750	500	1,250
Campbell, James A.		100	87	13	2,600	700	3,300
Jackson, Ben		184	94	90	2,400	400	2,800
Lyle, Robert		100	98	2	2,200	600	2,800
McKay, George H.		70	50	20	1,600	1,000	2,600
McLean, Thomas		525	325	200	6,000	1,000	7,000
McGregor, William		185	160	25	3,850	1,150	5,000
Oliver, George		100	80	20	2,400	800	3,200
Poole, T.A.[3]		300	215	85	4,200	1,000	5,200
Poole, Norman		200	100	100	4,200	1,000	5,200

AGE AND HOLDINGS, NORTH ELMSLEY TOWNSHIP

UFO Members

Total no. UFO members identified	12
Average acres owned	183.09
Average acres cleared	126.27
Average acres waste	56.82
Average value, land	$3,200.00
Average value, buildings	$850.00
Average total value	$4,050.00

General	
Total of sample surveyed	51
Average acres owned	166.31
Average acres cleared	125.49
Average acres waste	40.82
Average value, land	$3,146.08
Average value, buildings	$316.27
Average total value	$3,462.35

Source: Archives of Ontario, Assessment Rolls – North Elmsley Township, 1920

Notes:

1 Listed as a tenant. Figures not used in total calculations.

2 Listed as co-owner with William Cullen.

3 Listed as co-owner with A.H. Poole.

APPENDIX P

UFO Members Identified in Lanark Township, Lanark County

Name	*Age*	*Acres*	*Cleared*	*Nonprod*	*$Land*	*$Buildings*	*$Total*
Boyd, Hayes[1]	53	150	98	52	3,600	1,000	4,600
McKay, Alex[2]	61	550	362	188	2,300	1,100	3,400
Somerville, J.T.[3]	62	290	180	110	3,200	1,800	5,000
Campbell, Arch.	56	200	150	50	4,000	1,000	5,000
Thompson, George		100	80	20	900	1,000	1,900
McIlraith, George	65	100	45	55	1,400	600	2,000
Stewart, Robert[4]	28	367	205	162	3,700	1,000	4,700
Somerville, Wm[5]	32	846	150	696	3,000	1,000	4,000
McColl, Duncan[6]	52	0.5	0.5		200	700	900
Townend, J.B.[7]	37	0.5	0.5		200	800	1,000
Walters, William[8]	63	100	75	25	1,550	550	2,100

AGE AND HOLDINGS, LANARK TOWNSHIP

UFO Members

Total no. UFO members identified	11
Average age	50.90
Average acres owned	300.33
Average acres cleared	149.44
Average acres waste	150.89
Average value, land	$2,627.78
Average value, buildings	$1,005.56
Average total value	$3,633.34

General	
Total of sample surveyed	60
Average age	47.03
Average acres owned	228.85
Average acres cleared	123.10
Average acres waste	105.75
Average value, land	$2,352.75
Average value, buildings	$861.67
Average total value	$3,214.42

Source: Archives of Ontario, Assessment Rolls – Lanark Township, 1922

Notes:

1 Listed as co-owner with son, Franklin.

2 Listed as co-owner with son, Arthur.

3 Listed as co-owner with son, Roger.

4 Listed as co-owner with father, Arnold (age sixty), and brothers, Charles (twenty-six) and Thomas J. (twenty-four).

5 Listed as co-owner with father, David.

6 Occupation listed as Congregationalist clergyman. A supporter of the movement and quite possibly a member. Property statistics have not been included in final calculations.

7 Occupation listed as Presbyterian clergyman. A supporter of the movement and quite possibly a member. Property figures have not been included in final calculations.

8 Listed as co-owner with son, William, Jr.

APPENDIX Q

Platform of the Lanark Progressive Committee

PREAMBLE

We have in view a complete change in our present economic and social system. In this we recognize our solidarity the world over. As a means to this end and in order to meet the present pressing needs, we recommend the following platform:

1. Unemployment – State Insurance against Unemployment chargeable to Industry.
2. Public Ownership and Democratic Control of Public Utilities.
3. Electoral Reform – Proportional Representation. Names instead of Election Deposit. Extension of Voting Facilities.
4. Old Age Pensions and Health Disability Insurance.
5. Abolition of Non-elective Legislative Bodies.
6. International Disarmament.
7. Direct Legislation – Initiative, Referendum, Recall.
8. Enactment of Recommendations of Washington Labor Conference.
9. Repeal of Amendment to Immigration providing for Deportation of British Subjects.
10. Removal of Taxation on the Necessities of Life, Taxation of Land Values, and Abolition of Fiscal Legislation that Leads to Class Privileges.
11. Nationalization of the Banking System.
12. Capital Levy for Reduction of War Debt.
13. Full Publicity of all Election Expenses.

APPENDIX R

Federal and Provincial Election Results Lambton, Simcoe, and Lanark Counties

Candidate	*Party*	*Votes*	*%Votes*
LAMBTON EAST – FEDERAL RESULTS			
1921			
Fansher, Burt W.	Progr.	6,747	54.0
Armstrong, Joseph E.	Cons.	5,752	46.0
1925			
Armstrong, Joseph E.	Cons.	5,611	46.0
Fansher, Burt W.	Prog.	5,522	45.3
Stirrett, John R.	Lib.	1,061	8.7
1926			
Fansher, Burt W.	Prog.	6,891	52.1
Armstrong, Joseph E.	Cons.	6,340	47.9
1930			
Sproule, John T.	Cons.	6,209	49.4
Fansher, Burt W.	Prog.	6,196	49.3
Dunlop, Charles G.	Lib.	153	1.3
LAMBTON WEST – FEDERAL RESULTS			
1921			
LeSueur, Richard V.	Cons.	5,715	37.4
White, Robert J.	Progr.	4,958	32.5
Pardee, Frederick F. Lib.	Lib.	4,602	30.1
1925			
Goodison, William T.	Lib.	6,704	50.6
LeSueur, Richard V.	Cons.	6,535	49.4
1926			
Goodison, William T.	Lib.	7,551	50.5
LeSueur, Richard V.	Cons.	7,413	49.4

Candidate	*Party*	*Votes*	*%Votes*
1930			
Gray, Ross W.	Lib.	7,869	51.8
Haney, Wilfred S.	Cons.	7,314	48.2
LAMBTON EAST – PROVINCIAL RESULTS			
1919			
Oke, Leslie W.	UFO	4,575	53.1
Martyn, John B.	Cons.	2,161	25.1
McEachren, Duncan J.	Lib.	1,882	21.8
1923			
Oke, Leslie W.	UFO	3,224	44.3
Dawson, William R.	Cons.	2,776	38.0
Connelly, William G.	Lib.	1,291	17.7
1926			
Oke, Leslie W.	UFO	6,075	57.9
Sproule, John T.	Cons.	4,421	42.1
1929			
Fraleigh, Thomas H.	Cons.	4,632	43.9
Oke, Leslie W.	UFO	3,857	36.6
Eastman, Fred C.	Lib.	2,056	19.5
1934			
McVicar, Milton D.	Lib.	7,835	56.8
Fraleigh, Thomas H.	Cons.	5,429	39.4
Oke, Leslie W.	UFO	347	2.5
Fitzgerald, W.Y.	Ind.	183	1.3
LAMBTON WEST – PROVINCIAL RESULTS			
1919			
Webster, Jonah M.	UFO	6,081	40.4
Crawford, James S.	ILP	4,782	31.8
Gardiner, Peter	Cons.	4,180	27.8
1923			
Haney, Wilfred S.	Cons.	6,022	46.0
Webster, Jonah M.	UFO	3,903	29.8
Cook, Thomas H.	Lib.	3,179	24.2
1926			
Haney, Wilfred S.	Cons.	7,092	60.7
White, Robert J.	UFO	4,588	39.3
1929			
McMillen, Andrew R.	Cons.	5,724	55.0
Miler, William W.	Lib.	4,689	45.0
SIMCOE EAST – FEDERAL RESULTS			
1921			
Chew, Manley	Lib.	7,414	47.4
Raikes, Richard	Cons.	4,810	30.8
Swindle, Thomas F.	Prog.	3,414	21.8
1925			
Thompson, Alfred B.	Cons.	7,658	52.5
Chew, Manley	Lib.	6,929	47.5

Candidate	Party	Votes	%Votes
1926			
Thompson, Alfred B.	Cons.	7,994	51.0
Grant, Fred W.	Lib.	7,669	49.0
1930			
Thompson, Alfred B.	Cons.	7,974	51.1
McLean, George A.	Lib.	7,629	48.9
SIMCOE NORTH – FEDERAL RESULTS			
1921			
Ross, Thomas E.	Prog.	5,298	51.4
Currie, John A.	Cons.	4,489	43.5
Holden, William J.	Ind.	527	5.1
1925			
Boys, William A.	Cons.	6,885	52.2
Drury, Ernest C.	Prog.	6,295	47.8
1926			
Boys, William A.	Cons.	7,058	50.7
Drury, Ernest C.	Prog.	6,865	49.3
1930			
Simpson, John T.	Cons.	7,295	53.0
Drury, Ernest C.	Prog.	6,459	47.0
SIMCOE SOUTH – FEDERAL RESULTS			
1921			
Boys, William A.	Cons.	6,509	57.8
Jeffs, Compton B.	Prog.	4,758	42.2
Riding abolished 1924			
SIMCOE CENTRE – PROVINCIAL RESULTS			
1919			
Murdoch, Gilbert H.	UFO	5,234	57.9
Simpson, J.T.	Cons.	3,808	42.1
1923			
Wright, Charles E.	Cons.	3,535	39.8
Murdoch, Gilbert H.	UFO	3,006	33.9
Simpson, Leonard J.	Lib.	2,332	26.3
1926			
Wright, Charles E.	Cons.	5,313	50.9
Todd, Ebenezer	Lib.	5,120	49.1
1929			
Simpson, Leonard J.	Lib.	5,199	51.1
Forgie, John	Cons.	4,979	48.9
SIMCOE EAST – PROVINCIAL RESULTS			
1919			
Johnston, John B.	UFO/LIB	5,063	40.8
Hartt, James I.	Cons.	4,580	36.9
Anderson, Duncan C.	Ind.	2,773	22.3
1923			
Finlayson, William	Cons.	5,692	57.6
Johnston, John B.	UFO	4,194	42.4

Candidate	*Party*	*Votes*	*%Votes*
1926			
Finlayson, William	Cons.	7,312	55.8
Ross, Thomas E.	Lib.Prog.	5,782	44.2
1929			
Finlayson, William	Cons.	7,980	61.6
Harvey, James G.	Lib.	4,969	38.4
SOUTH SIMCOE – PROVINCIAL RESULTS			
1919			
Evans, Edgar J.	UFO	2,927	53.7
Ferguson, Alexander	Cons.	2,526	46.3
1923			
Rowe, William E.	Cons.	3,016	55.9
Evans, Edgar J.	UFO	2,381	44.1
Riding abolished 1926			
SIMCOE SOUTHWEST – PROVINCIAL RESULTS			
1926			
Mitchell, John H.	Lib./Prog.	5,779	52.0
Jamieson, James E.	Cons.	5,327	48.0
1929			
Jamieson, James E.	Cons.	6,213	54.4
Mitchell, John H.	Lib.	5,200	45.6
SIMCOE WEST – PROVINCIAL RESULTS			
1919			
Allan, William T.	Cons.	4,491	55.5
Baker, Richard	UFO	3,606	44.5
1923			
Jamieson, James E.	Cons.	3,610	56.8
Baker, Richard	UFO	2,030	31.9
Currie, High A.	Lib.	634	10.0
Carmichael, William	Ind.	84	1.3
Riding abolished 1926			
LANARK – FEDERAL RESULTS			
1921			
Stewart, John A.	Cons.	9,250	57.7
Anderson, Robert M.	Prog.	6,615	41.3
Ferguson, William G.	Lab.	158	1.0
1922			
Preston, Richard F.	Cons.	8,497	54.7
Findlay, David	Lib.	7,048	45.3
1925			
Preston, Richard F.	Cons.	7,620	63.3
Gemmell, Duncan H.	Ind.	4,416	36.7
1926			
Preston, Richard F.	Cons.	8,122	62.3
Buchanan, George W.	Prog.	4,908	37.7
1929			
Murphy, William S.	Ind. Cons.	7,174	55.8
Thompson, Thomas A.	Cons.	5,682	44.2

Candidate	*Party*	*Votes*	*%Votes*
1930			
Thompson, Thomas A.	Cons.	7,064	42.1
Soper, Bert H.	Lib.	5,699	34.0
Murphy, William S.	Ind. Cons.	3,937	23.5
Low, Mildred A.	Ind.	75	0.4
LANARK NORTH – PROVINCIAL RESULTS			
1919			
McCreary, Hiram	UFO	2,881	40.9
Preston, Richard F.	Cons.	2,798	39.7
Forbes, Christopher	Lib.	1,272	19.4
1923			
Thompson, Thomas A.	Cons.	3,339	54.3
McCreary, Hiram	UFO	2,808	45.7
1926			
Thompson, Thomas A.	Cons.	3,589	51.7
Caldwell, William R.	Lib./Prog.	3,353	48.3
1929			
Craig, John A.	Cons.	4,038	56.6
Downing, Albert	Lib.	3,091	43.4
LANARK SOUTH – PROVINCIAL RESULTS			
1919			
Johnson, William I.	UFO	3,872	48.2
Gould, James S.	Cons.	3,069	38.2
Grant, Richard	ILP	1,096	13.6
1923			
Stedman, Egerton R.	Cons.	3,874	57.3
Johnson, William I.	UFO	2,891	42.8
1926			
Stedman, Egerton R.	Cons.	3,874	56.8
Wickware, Ernest H.	Lib.	2,947	43.2
1929			
Anderson, James A.	Cons.	4,308	57.4
Anderson, James E.	Lib.	3,197	42.6

Source: Centennial Edition of a History of the Electoral Districts, Legislatures and Ministries of the Province of Ontario 1867–1967; History of the Federal Electoral Ridings 1867–1992

APPENDIX S

The Farmers' Platform

Drafted by the Canadian Council of Agriculture 29 November 1918 and accepted by the member organizations, 1919.

1 A League of Nations as an international organization to give permanence to the world's peace by removing old causes of conflict.

2 We believe that the further development of the British Empire should be sought along the lines of partnership between nations free and equal, under the present governmental system of British constitutional authority. We are strongly opposed to any attempt to centralize imperial control. Any attempt to set up an independent authority with power to bind the Dominions, whether this authority be termed parliament, council or cabinet, would hamper the growth of responsible and informed democracy in the Dominions.

THE TARIFF

3 Whereas Canada is now confronted with a huge national war debt and other greatly increased financial obligations, which can be most readily and effectively reduced by the development of our natural resources, chief of which is agricultural lands;

And whereas it is desirable that an agricultural career should be made attractive to our returned soldiers and the large anticipated immigration, and owing to the fact that this can best be accomplished by the development of a national policy which will reduce to a minimum the cost of living and the cost of production;

And whereas the war has revealed the amazing financial strength of Great Britain, which has enabled her to finance, not only her own part in the struggle, but also to assist in financing her Allies to the extent of hundreds of millions of pounds, this enviable position being due to the free trade policy which has enabled her to draw her supplies freely from every quarter of the globe and consequently to undersell her competitors on the world's market, and because this policy has not only been profitable to Great Britain, but has greatly strengthened the bonds of Empire by facilitating trade between the Motherland and her overseas Dominions – we believe that the best interests of the Empire and of Canada would be served by reciprocal action on the part of Canada through gradual reduction of the tariff on British imports, having for its objects closer union and a better understanding between Canada and the Motherland and at the same time bring about a greater reduction in the cost of living to our Canadian people;

FOSTERS COMBINES

And whereas the Protective Tariff has fostered combines, trusts and "gentlemen's agreements" in almost every line of Canadian industrial enterprise, by means of which the people of Canada – both urban and rural – have been shamefully exploited through the elimination of competition, the ruination of many of our smaller industries and the advancement of prices on practically all manufactured goods to the full extent permitted by the tariff;

And whereas agriculture – the basic industry upon which the success of all our other industries primarily depends – is unduly handicapped throughout Canada as shown by the declining rural population in both Eastern and Western Canada, due largely to the greatly increased cost of agricultural implements and machinery, clothing, boots and shoes, building material and practically everything the farmer has to buy, caused by the Protective Tariff, so that it is becoming impossible for farmers generally, under normal conditions, to carry on farming operations profitably;

And whereas the Protective Tariff is the most wasteful and costly method ever designed for raising national revenue, because for every dollar obtained thereby for the public treasury at least three dollars pass into the pockets of the protected interests thereby building up a privileged class at the expense of the masses, thus making the rich richer and the poor poorer;

And whereas the Protective Tariff has been and is a chief corrupting influence in our national life because the protected interests, in order to maintain their unjust privileges, have contributed lavishly to political and campaign funds, thus encouraging both political parties to look to them for support, thereby lowering the standard of public morality;

DEFINITE TARIFF DEMANDS

Therefore be it resolved that the Canadian Council of Agriculture, representing the organized farmers of Canada, urges that, as a means of remedying

these evils and bringing about much-needed social and economic reforms, our tariff laws should be amended as follows:
(a) By an immediate and substantial all-round reduction of the customs tariff.
(b) By reducing the customs duty on goods imported from Great Britain to one-half the rates charged under the general tariff, and that further gradual, uniform reductions be made in the remaining tariff on British imports that will ensure complete Free Trade between Great Britain and Canada in five years.
(c) By endeavoring to secure unrestricted trade in natural products with the United States along the lines of the Reciprocity Agreement of 1911.
(d) By placing all foodstuffs on the free list.
(e) That agricultural implements, farm and household machinery, vehicles, fertilizers, coal, lumber, cement, gasoline, illuminating, fuel and lubricating oils be placed on the free list, and that all raw materials and machinery used in their manufacture also be placed on the free list.
(f) That all tariff concessions granted to other countries be immediately extended to Great Britain.
(g) That all corporations engaged in the manufacture of products protected by the customs tariff be obliged to publish annually comprehensive and accurate statements of their earnings.
(h) That every claim for tariff protection by any industry should be heard publicly before a special committee of parliament.

TAXATION PROPOSALS

4 As these tariff reductions may very considerably reduce the national revenue from that source, the Canadian Council of Agriculture would recommend that, in order to provide the necessary additional revenue for carrying on the government of the country and for the bearing of the cost of the war, direct taxation be imposed in the following manner:
(a) By a direct tax on unimproved land values, including all natural resources.
(b) By a graduated personal income tax.
(c) By a graduated inheritance tax on large estates.
(d) By a graduated income tax on the profits of corporations.
(e) That in levying and collecting the business profits tax the Dominion Government should insist that it be absolutely upon the basis of the actual cash invested in the business and that no considerations be allowed for what is popularly known as watered stock.
(f) That no more natural resources be alienated from the Crown, but brought into use only under short-term leases, in which the interests of the public shall be properly safeguarded, such leases to be granted only by public auction.

THE RETURNED SOLDIERS

5 With regard to the returned soldier we urge:
(a) That it is the recognized duty of Canada to exercise all due diligence for the future well-being of the returned soldier and his dependents.
(b) That demobilization should take place only after return to Canada.
(c) That first selection for return and demobilization should be made in the order of length of service of those who have definite occupation awaiting them or have assumed other means of support, preference being given first to married men and then to the relative need of industries, with care to insure so far as possible the discharge of farmers in time for the opening of spring work upon the land.
(d) That general demobilization should be gradual, aiming at the discharge of men only as it is found possible to secure steady employment.
(e) It is highly desirable that if physically fit discharged men should endeavor to return to their former occupations, all employers should be urged to reinstate such men in their former positions wherever possible.
(f) That vocational training should be provided for those who while in the service have become unfitted for their former occupation.
(g) That provision should be made for insurance at the public expense of unpensioned men who have become undesirable insurance risks while in the service.
(h) The facilities should be provided at the public expense that will enable returned soldiers to settle upon farming land when by training or experience they are qualified to do so.
6 We recognize the very serious problem confronting labor in urban industry resulting from the cessation of war, and we urge that every means, economically feasible and practicable, should be used by federal, provincial and municipal authorities in relieving unemployment in the cities and towns; and, further, recommend the adoption of the principle of cooperation as the guiding spirit in the future relations between employer and employees – between capital and labor.

LAND SETTLEMENT

7 A land settlement scheme on a regulating influence in the selling price of land. Owners of idle areas should be obliged to file a selling price on their lands, that price also to be regarded as an assessable value for purposes of taxation.
8 Extension of cooperative agencies in agriculture to cover the whole field of marketing, including arrangements with consumers' societies for the supplying of foodstuffs at the lowest rates and with the minimum of middleman handling.

9 Public ownership and control of railway, water, and aerial transportation, telephone, telegraph, and express systems, all projects in the development of natural power, and of the coal mining industry.

OTHER DEMOCRATIC REFORMS

10 To bring about a greater measure of democracy in government, we recommend:

(a) That the new Dominion Election Act shall be based upon the principle of establishing the federal electorate on the provincial franchise.

(b) The discontinuance of the practice of conferring titles upon citizens of Canada.

(c) The reform of the federal senate.

(d) An immediate check upon the growth of government by order-in-council, and increased responsibility of individual members of Parliament in all legislation.

(e) The complete abolition of the patronage system.

(f) The publication of contributions and expenditures both before and after electoral campaigns.

(g) The removal of press censorship upon the restoration of peace and the immediate restoration of the rights of free speech.

(h) The setting forth by daily newspapers and periodical publications, of the facts of their ownership and control.

(i) Proportional representation.

(j) The establishment of measures of direct legislation through the initiative, referendum, and recall.

(k) The opening of seats in Parliament to women on the same terms as men.

(l) Prohibition of the manufacture, importation, and sale of intoxicating liquors as beverages in Canada.

APPENDIX T

Central UFWO Suggested Fall Meeting Program – 1925

SEPTEMBER

Community singing; prayer
Business – Minutes, treasurer's report; business arising out of minutes, reports from convenors on the various departments of work, correspondence, new business
Roll Call – "A plant or shrub I'd like to have"
Current events; instrumental; papers, "Civics"; solo; paper, "A well-balanced meal"; closing remarks

OCTOBER

Community singing; opening exercises; business, study of resolutions to be submitted to the Head Office
Roll call – "A favorite book"
Current events; solo; paper, "How we can help ourselves in agriculture"; instrumental; paper, "Importance of rest and recreation for women"
Closing remarks

NOVEMBER

Commuity singing; opening exercises; business; study of resolutions for annual meeting
Roll call – "Hints for club improvement"

Current events; music; paper, "World peace"; reports from officers and convenors on year's work
Election of officers
Closing remarks

DECEMBER

Community singing; opening exercises; business; study of remaining resolutions for annual meeting
Roll call – "Christmas suggestions"
Current events; music; paper, "Royal Winter Fair in Toronto"; instrumental; paper, "Christmas"
Closing remarks

Source: SUN, 27 August 1925, 8.

Notes

CHAPTER ONE

1 Chomsky, *Towards a New Cold War*, 9.
2 Despite the lack of local records, Lanark was chosen because it had four towns in which weekly newspapers were published.
3 An average shift of three percent in the vote in twenty other ridings would have resulted in Patron victories; Shortt, "Social Change and Political Crisis in Rural Ontario," 222.
4 NA, MG 27 III D 3, "Memoirs of J.J. Morrison," 14–16; Badgley, "The Social and Political Thought of the Farmers' Institute," 105–6.
5 Wood, *A History of Farmers' Movements in Canada*, 311–13.
6 This study also seeks to add to the sparse literature on the Ontario dimensions of the Progressive party. Usually, the Progressive experience in the province is summed up in a few paragraphs. See W.L. Morton, *The Progressive Party in Canada*; Thompson with Seager, *Canada 1921–39*, 14–37.
7 Wood, *A History*, 273–84; MacLeod, "The United Farmer Movement," 32–75.
8 Wood, *A History*; Hanam, *Pulling Together for 25 Years*; Staples, *The Challenge of Agriculture*; Irvine, *The Farmers in Politics*.
9 Drury, *Farmer Premier*; Good, *Farmer Citizen*.
10 Hillman, "J.J. Morrison"; Johnston, *E.C. Drury*; Wylie, "Direct Democrat; Thomas, "The Ideas of W.C. Good"; Pennington, *Agnes Macphail*; Crowley, *Agnes Macphail and the Politics of Equality*.

11 MacLeod, "The United Farmer Movement"; Bristow, "Agrarian Interest"; J.D. Hoffman, "Farmer-Labour Government in Ontario; Trowbridge, "Wartime Rural Discontent."
12 Young, "Conscription." See also Van Loon, "Political Thought." Van Loon's interpretation differs from others in that he sees the movement as being more than a group of disenchanted Liberals (pp. 68–9). At the same time, however, he argues that its members were "afraid of change" (p. 117).
13 Brown, "The Broadening Out Controversy"; Crowley, "The New Canada Movement"; Griezic, "An Introduction," xiii. Griezic points to conflicting views within the movement, and to the decision to focus on cooperatives rather than politics as other reasons for its decline (pp. xii, xxi).
14 See, for example, Tennyson, "The Ontario General Election of 1919." One exception is Griezic's "'Power to the People.'" Griezic, however, concentrates on the by-election and does not address tensions between the central UFO and the local clubs. See, as well, Hoffman, "Intra-Party Democracy: A Case Study." Despite not going into detail about what occurred at the local level, Hoffman argues that the "parliamentary" side of the movement felt forced "to compromise with a more complex political reality" and thus departed "from the … doctrine of the grass-roots organization" (p. 233). See also Anderson et al., *A Political History.* Most of this unfocused book concentrates on the UFO's leadership.
15 See Oliver's essay collection, *Public & Private Persons*, especially chapters 1 and 2, and his *G. Howard Ferguson*, especially chapters 5 to 8. See also MacDonald, "Ontario's Political Culture."
16 On the formative years of conservatism in Ontario, Wise remains the authority. See, for example, his "Upper Canada and the Conservative Tradition," and "The Ontario Political Culture."
17 See Desmond Morton, "*Sic Permanet.*"
18 Evans, *Sir Oliver Mowat*; Charles W. Humphries, "*Honest Enough to Be Bold*".
19 For a critique of the brokerage model, see Horowitz, "Toward the Democratic Class Struggle." Although works on the Ontario political culture occasionally refer to non-mainstream groups, they are not taken seriously, even when they manage to exert political influence. See, for example, MacDonald's treatment of the Knights of Labour, the POI and the UFO in "Ontario's Political Culture," 310, 313.
20 Noel, *Patrons, Clients, Brokers*, 14–15.
21 Ibid., 299–306, 317.
22 Gagan, "Writing the History of Ontario in the 1980s," 166, 179.

23 Bothwell, *A Short History of Ontario*, 120–3. Bothwell concentrates exclusively on the period when the UFO formed the government and does not mention the role Ontario Progressives played in the 1921 federal election. The UFO fares slightly better in Randall White's history of the province, *Ontario 1610–1985*. He argues (p. 213) that the 1919 election victory was the "one point in Ontario history ... when genuinely spontaneous mass protest might be said to have played a decisive role in regional politics."

24 See Bliss, *Northern Enterprise*. Though he claims that agriculture, "Canada's most important single industry," has been "traditionally slighted in the writing of business history" (p. 18), Bliss largely ignores it too. On the dismissive side are scholars such as Young ("The Progressives"), who argue that the movement was an ill-conceived blip on an otherwise calm political sea that foundered on its misunderstanding of who the real enemies were and on the contradictions in its platform. Young's thoughts are deemed to be competent enough to be included in a text for undergraduates.

25 Sandwell, "Rural Reconstruction," 7–10. She notes that in recent studies there is a tendency to abandon the distinction between "traditional peasantry" and "modern agrarian capitalists" and instead to see late-nineteenth-century farmers as "petty producers who grew much of their own food, and participated in commercial markets," employing Allan Kulikoff's words. Ibid., 15. See Kulikoff, "The Transition to Capitalism in Early America."

26 Sandwell, "Rural Reconstruction," 16, 29, 32.

27 Sandwell raises interesting points, but there are problems in her assessment. She believes, for example, that there is much to be said about recent work that explores the "gendered relations of power in society." There is no doubt that these studies are valuable, but one wonders why Sandwell does not mention other equally pervasive forms of power. On this point I agree with Bryan D. Palmer who, in writing on the working class, argues that it "cannot be conceived or understood as *un*gendered, but neither can history be reduced ... *only* to gender and its oppressions." Palmer, *Working Class Experience*, 24–5.

28 Thompson, "Writing about Rural Life," 108–9. For those who feel Thompson's caricature is too strong, see Ghorayshi, "Canadian Agriculture."

29 See, for example, Voisey, *Vulcan*; Samson, ed., *Contested Countryside*.

30 See, for example, Elliott, *Irish Migrants in the Canadas*. Other communities of interest, such as producers of specific commodities, have also recently received some attention. See, for example, Menzies, *By the Labour of Their Hands*.

31 See, for example, Lockwood, *Beckwith*; Graham, *Greenbank*. Although he has produced a competent study, Graham shares with other local historians an unwillingness to explore the role of politics in rural communities. While he goes into the twentieth century with some of his subjects, there is no mention of the UFO. This is a curious oversight, especially since the second UFO candidate to be elected (John Widdifield) represented the riding in which Greenbank is located.

32 See, for example, McInnis, *Perspectives on Ontario Agriculture 1815–1930*; McCalla, *Planting the Province*.

33 See, for example, Kechnie, "The United Farm Women of Ontario"; Rankin, "The Politicization of Ontario Farm Women"; Ambrose, *For Home and Country*.

34 A good example of this genre is MacGillivray, *The Slopes of the Andes*. In addition, two recent works attempt to describe aspects of Ontario's history in terms of residual rural culture. In his study of the margarine debate in Canada (which devotes considerable space to Ontario), Heick argues that it took Canada longer than other countries to legalize the sale of margarine because of the slow transformation "from a rural society to one with predominantly urban perspectives" (*A Propensity to Protect*, vii). Why butter was the symbol of rural values is not explained. In a study of the Province of Ontario Savings Office (created by the Drury government in 1922), White attributes its success in no small part to the folksy, friendly smalltown service it offered to its clients ("The Province of Ontario Savings Office, 1922–1990").

35 Richards, "The New Populism," 263–4. See also his "Populism: A Qualified Defence"; and Crowley, *Macphail*, in which populism is defined (p. 26) as a "conception of the political process that seeks to overcome barriers separating those who govern from the people governed ... [W]hen they are successful at the polls, the complexity and variety of political issues creates inevitable tensions between those within the populist movement who comprise its base and those who get elected."

36 See Kealey and Palmer, *Dreaming of What Might Be*, in which they wrote that the 1890s was a period of agrarian revolt that found its voice in the POI. "Within this agrarian upheaval flourished the same kinds of rhetoric and analysis used by the [Knights of Labor] throughout the 1880s. This should not surprise us, for the Patrons contained an active contingent of former Knights ... who sought to create a farmer-labour alliance capable of reforming society from top to bottom" (pp. 387–8). What is not mentioned is how the POI could so easily accommodate the ideas and rhetoric of the Knights of Labor; doing so would mean that the authors would have to allow that the

social critique developed in their industrial proletariat could be pursued as strongly by "petit bourgeois" farmers.

37 The dismissal of farmers often takes the form of omission. See, for example, Kealey, "State Repression of Labour."

38 Hann, *Farmers Confront Industrialism*, 2, 9. Hann also provides a good account of the "agrarian myth," the farmers' belief that agriculture was a more noble calling than other occupations because people could not survive without the food that farmers produced, and because farmers, in working the land, were closer to the natural world and thus had a better appreciation of how nature unfolded (pp. 9–13).

39 Wylie, 1–4. For a similar argument regarding Prairie populists, see Conway, "'To Seek a Goodly Heritage'."

40 Wylie, "Direct Democrat," 5–7.

41 Wylie also suggests that the movement died as a result of the split between crypto-liberals and direct democrats. This was a component in the movement's decline, but not the only one. By focusing on the split as it manifested itself within the movement's leadership, Wylie fails to explain, as does everyone who subscribes to this view, how the rank and file reacted to the split.

42 A good introduction to anarchism and the several streams within it is Woodcock, *Anarchism*. See also Guerin, *Anarchism*; Bakunin, *Bakunin on Anarchy*.

43 Rocker, *Anarchosyndicalism*, 31, cited in Chomsky, *For Reasons of State*, 370. See also Harrison, *The Modern State*, especially 13–5; Clark, *The Anarchist Moment*, especially 19–32. Chomsky defines anarchism as "a historical tendency ... of thought and action, which has many different ways of developing and progressing and which ... will continue as a permanent strand of human history." *The Chomsky Reader*, 29.

44 See Morris, *Bakunin*, especially 81, 114–15; Craig, *The Nature of Cooperation*.

45 For an attempt to go beyond these positions, see Brown, *The Politics of Individualism*.

46 "Anarchism is a doctrine that poses a criticism of existing society; a view of a desirable future; and a means of passing from one to the other." Woodcock, *Anarchism*, 7.

47 Goodwyn, *Democratic Promise*, *The Populist Moment*.

48 Goodwyn, *Populist Moment*, xix.

49 Ibid., xvii, xviii.

50 Ibid., xi.

51 In *Dreaming*, Kealey and Palmer acknowledge their debt to Goodwyn for providing the analytical framework that allowed them to see the Knights of Labor as "a movement culture of alternative, opposition, and potential" (p. 17).

52 Naylor, *The New Democracy*, 3.

53 See, for example, his account of the workers' call for reform of the electoral system. Although it may appear naive to present-day historians, Naylor believes that "it is important to recognize that, when combined with the democratic rhetoric that accompanied the war and the growing disregard for democratic procedure by governments, the result was explosive. Labour's call for a wider, participatory democracy threatened a greater impact than the narrowly electoral character of its demands." Ibid., 84.

54 Discussing workers' nativism, Naylor notes that they faced a barrage of anti-alien propaganda: "The press, politicians, and businessmen vilified 'the Hun' during the war and ... scapegoated immigrants in a post-war 'red scare.'" Ibid., 103. Such comments should be read in conjunction with Keshen, "All the News That Was Fit to Print." Keshen largely ignores those who opposed such censorship at the time, even though farmers had much to say about the subject, despite the legal restrictions placed upon their right to express such views. Second, he argues that Chief Censor Chambers's tactics "may have appeared unnecessarily tough" but then states that his actions merely "mirrored a society frantic in its fear of enemy aliens and the rise of the left" (p. 343). Even if Keshen had proved that this was the case, the issue seems to be a horse-and-cart one: to what extent was this public hysteria (such as it was) the result of press propaganda that related countless "Hun atrocities" and disparaged the left at every opportunity?

55 Naylor, *The New Democracy*, 122.

56 Ibid., 252.

57 Laycock is not the only political scientist to see the agrarian protest as being more than the efforts of cranky independent commodity producers. Reg Whitaker ("Images of the State in Canada," 51) argues that the political theory that agrarians developed in the early 1920s and just before "was in fact radically divergent from that of the dominant forces in Canadian life, which had used representative and parliamentary governmental institutions to divide and mystify the subordinate classes." See also his introduction to Irvine's study of agrarian protest, in which he notes that "the farmers' movement did involve the vision of a better world that was indeed radical by the standards of the time." In Irvine, *The Farmers*, xi.

58 Laycock, *Populism and Democratic Thought*, 3, 4. See also p. 267, where he notes that, although the movement had a class-based character, "this class basis did not exclusively define its extended conceptions of democracy. Class attachments do not necessarily produce all-embracing class logics."

59 Ibid., 29.

60 Ibid., 12.

61 Laycock also states that another conceptualization of the state is possible within a populist paradigm. In almost all populist movements, he notes, there is a hostility towards the state because it is seen as a key player in maintaining inequitable social relations: "Those populisms that advocate a socialization of power develop a vague conception of the state as the political embodiment of domination *per se*. In this sense, the logical extension of the democratic idea is the elimination of the state itself. Populism here crosses paths with anarchism." Ibid., 12.

62 Laycock also downplays the power of the "official" culture to modify the response of populists, although he is more attentive to this issue in his studies of present-day politics. See his "Democracy and Cooperative Practice."

63 Laycock, *Populism and Democratif Thought*, 3.

64 Ibid., 26.

65 Leigh, *Monty Leigh Remembers*, 18.

66 There were some, no doubt, who used the UFO out of political ambition. Mitch Hepburn, for example, was once a UFO supporter. Saywell, *'Just Call me Mitch'*, 12–23.

CHAPTER TWO

1 See, for example, Wylie, "Direct Democrat," 4–9. Johnston, in "'A Motley Crowd,'" argues the contrary and claims that the UFO consisted of different types of farmers (i.e., dairymen, fruit growers, cattlemen, etc.).

2 The East Lambton UFO Political Association minutes were also located, but they do not reveal the membership in that locality, save for a few executive members.

3 Elford, *Canada West's Last Frontier*, 3.

4 Ibid., 1–7, 130–40, 146–54. See also Lauriston, *Lambton's Hundred Years*, 8–23, 290–316; and Belden & Co., *Belden's Illustrated Historical Atlas of the County of Lambton*, 3–20. For a succinct history of Lambton, see Mika and Mika, *Places in Ontario*, 486–8.

5 See Angus, *A Deo Victoria.*

6 For an overview of late-nineteenth-century Simcoe County, see Craig, *Simcoe County*, 1–24. See also Hunter, *A History of Simcoe County*, and Belden & Co., *Historical Atlas of Simcoe County*, 3–20. See also Mika and Mika, *Places in Ontario*, 395–8.

7 By the late nineteenth century Lanark was one of the province's leading textile producers. See Bland, "The Location of Manufacturing." See also Price, "The Changing Geography."

8 On lumbering in Lanark, see Brown, *Lanark Legacy*, 228–51. On iron foundries and other industries, see Bennett, *In Search of Lanark*,

especially her sections on Carleton Place and Smiths Falls. See also Mika and Mika, *Places in Ontario*, 490–2; and Belden & Co., *Historical Atlas of Lanark and Renfrew Counties*, 13–23. For a listing of manufacturers in Perth, see Turner (with Stewart), *Perth*, 123–5.

9 Reeds, "The Environment," 9–10. The work of L.G. Reeds himself also points to these conclusions; see ibid., 10–22.

10 Ontario, Department of Agriculture and Food, *Origins*, 12–20, 45–54.

11 Census of Canada, 1921. Unless otherwise noted, data in this section is from the Census of Canada of either 1911, 1921, or 1931.

12 Census of Canada, 1921. The figures presented here differ from those in Table 1, which were provincially generated, possibly due to the difference in terminology ("farmland" as opposed to "rural" land). Little work has been undertaken to explore the relationship between farm size and prosperity. Even William Marr, who has produced excellent work on farm size in nineteenth-century Ontario, is more concerned about how farm size aided in determining how resources were allocated and in what types of crops were grown. Marr argues that so few studies exist because such work involves tracking down individual farms through manuscript census returns, a laborious process. For my purposes, studying the relationship between farm size and prosperity would also necessitate a literal tracking down of each farm used in the study in order to determine the quality of soil, number of hills and rocks, etc. See Marr's "Did Farm Size Matter?"

13 Census of Canada 1931, 416–17, 421.

14 See Ontario, Department of Agriculture, *Annual Report*, 37.

15 McInnis, 93–4.

16 In the assessments, non-productive land was divided into three categories: woodland; slash; and swamp, marsh, or wasteland. In the following calculations, these categories were combined.

17 Assessment rolls were used in two important Ontario historical demographic studies, Katz's *The People of Hamilton, Canada West*; and Gagan's *Hopeful Travellers*. Gagan notes that assessment records for Peel County for the period in question were badly preserved, and were thus of little value (9). Katz had a different experience and argues that assessment rolls can "supplement the manuscript census with detailed economic information" (18). For an example of what a survey of assessment rolls can produce, see Russell, "Upper Canada."

18 The decision was made to use samples rather than to total all of the farm figures entered for township farmers after it was discovered that many irregularities exist in the assessment rolls. For example, some farms were run as estate farms or they were operated as corporations. Such farms were not included in the sample group. In addition, there

were many examples of rather large farms owned by groups of people. These too have been eliminated, as have other atypical entries that might skew the results of the calculations.

19 As noted in Appendix K, part of the reason for the difference in average values is the fairly high values for farm buildings owned by UFO members. The values for land were very close.

20 In the case of the Oro Station UFO, the years 1917–30 were used for locating executive members. For the Edenvale UFO, the years 1920–34 were employed; and for Rugby, the years 1920–30 were used.

As Table 16 shows, a good percentage of members served in some capacity during the period in question, which speaks well for the democratic spirit in each club, though only those who served on the executive, and not those appointed to committees or named as delegates to conventions, are enumerated in the tables.

Executive and non-executive members from these clubs were compared in a manner similar to that of farmers in the appendices (Table 17). In the case of the Edenvale club, executive members appear to have been much better off than non-executive members. A similar pattern – although less marked – exists in the case of the Oro Station club. In Rugby, however, one finds the situation reversed, with executive members less well off than their non-executive counterparts. It is not known why Rugby exhibited this characteristic, and there is no way of determining whether it was merely an anomaly. If the Rugby situation was atypical, and if the Edenvale and Oro Station clubs can be taken as representative of UFO clubs throughout the province, then, if anything, the figures for assets owned by Lambton and Lanark UFO members might actually be inflated. In other words, the economic status of rank-and-file members (as opposed to those who served on club executives) might be demonstrably lower than indicated.

21 See, for example, Clarke et al., *Political Choice in Canada*, 381–8; Van Loon, "Political Participation in Canada," 389; Milbrath, *Political Participation*, 134–5.

22 The religious affiliations of Lambton residents were not recorded by assessors at that time.

23 Allen, *The Social Passion*, 15–6. The Social Gospel alludes to a movement that attempted to apply Christian principles as a means of remedying social ills. Allen's work on the movement remains the standard account.

24 John S. Moir, *Enduring Witness*, 209–22.

25 Clifford, *The Resistance to Church Union*, 183–4. Allen (*The Social Passion*, 202, 211–12) documents a strong link between the Social Gospel and the Progressives.

26 R.S. Pennefather argues that the Orange Lodge played a large role in the 1923 defeat of the UFO/ILP government. See "The Orange Order," 169.

27 The number of marks on the map does not correspond to the number of farmers identified because farmers' sons sharing land with their fathers have not been noted.

28 Only twenty-four out of forty-three total members have been plotted for the Rugby UFO, and only thirteen out of thirty-seven total members for the Edenvale UFO.

29 On these surveys, see Wylie *Direct Democrat*, 296–9; *SUN*, 15 January 1919, 4; 22 January 1919, 4.

30 Most farmers, however, did not pay rent and had lower food and heating costs.

CHAPTER THREE

1 *SFRN*, 6 December 1921, 5.

2 Goodwyn, *Populist Moment*, xxi.

3 *SUN*, 21 August 1920, 7.

4 *SFRN*, 1 September 1921, 1, 7.

5 It will be noted that there are frequent references to UFO leaders in this chapter. The decision to write with such an emphasis was made reluctantly, but given the scarcity of local sources, there was no other recourse. Quite simply, the words of the movement's leadership were recorded much more frequently than those of rank-and-file members. Their speeches are not without merit and have utility since they convey the ideas that were current at the time, the ideals to which average members were exposed. It may be difficult to determine the extent to which UFO leaders were sincere in their opinions, but it is irrefutable that their speeches had a profound impact on local farmers. Through its pronouncements, the UFO leadership helped to provide rank-and-file members with the means to express their discontent and, by 1921, these same members – including Lambton, Simcoe, and Lanark farmers – were using the movement to advance their social critique.

6 There is scant pre-1919 documentation relating to UFO clubs. Wylie notes that until 1916 the UFO "barely existed as an entity independent of the (UFCC). What actions it did take were of a lobby character" (Wylie, *Direct Democrat*, 327–8). Before it contested the 1919 election, local newspaper coverage of the UFO was sparse, and local records, where they exist, consist largely of membership lists and brief meeting minutes. Hence little attention is paid here to the movement

before 1919, aside from rural/urban tensions and the UFO's stand on conscription – two issues that were covered fairly regularly by the press.

7 *OT*, 24 November 1921, 4.

8 Wood, 276, 282–3; *CB*, 19 February 1920, 2.

9 In 1921 some 295,000 people were engaged in agriculture in Ontario. If there were 60,000 UFO members at that time, then members equalled roughly twenty percent of the farming population. Drummond, *Progress without Planning*, 362.

10 *SUN*, 7 July 1915, 5; 4 July 1917, 5; 2 April 1919, 10; 26 November 1919, 10. See also *FS*, 22 June 1916, 2. New clubs were formed in Lambton as late as 1921. *SUN*, 12 January 1921, 7; *FS*, 20 June 1921, 1.

11 *OT*, 14 February 1918, 2. Over sixty farmers joined at the inaugural meeting of the Loretta club. *SUN*, 12 June 1918, 3. On the other clubs, see *CE*, 8 August 1918, 1; *SUN*, 2 April 1919, 3. The Edenvale club affiliated with the UFO in April 1919. Simcoe County Archives, Edenvale Minutes, 9 April 1919. Distinguishing between UFO and Farmers' clubs downplays the UFO's influence in a locale. The Oro Station Farmers' Club, before affiliating with the UFO, sent delegates to local UFO conventions and formally supported UFO candidates in the 1919 provincial election. Simcoe County Archives, Oro Minutes, 22 July 1919, 2 October 1919, 6 November 1919.

12 *CE*, 2 October 1919, 4.

13 *SUN*, 2 April 1919, 10; 11 June 1919, 10. A provincial official noted that the average attendance for Carleton Place UFO club meetings was over seventy in 1921. AO, RG 16, 16. The Ramsay Farmers' Club formally affiliated with the UFO in early 1919, although it had ties before that time. *AG*, 21 March 1919, 4.

14 *SFRN*, 16 September 1919, 1; *CPH*, 17 September 1919, 1; *AG*, 19 September 1919, 1. It was also noted at the time that there were 3,873 enumerated rural voters. If these figures are accurate, then a significant percentage of households in Lanark had more than passing familiarity with the UFO. See also *PC*, 30 January 1920, 3.

15 Accounts of "monster" picnics abound. See, for example, *CE*, 10 July 1919, 1, where it was noted that a UFO picnic in Simcoe that month attracted three thousand people. In 1917 the Forest United Farmers' Association picnic drew roughly five thousand people. *FS*, 5 July 1917, 2. The Ramsay Farmers' Club picnic in 1921 was attended by some two thousand people. *AG*, 17 June 1921, 1.

16 *CB*, 3 April 1919, 3.

17 *FS*, 14 March 1918, 6. Simcoe's Stroud UFO organized a curling club. *CSN*, 22 January 1921, 4. In many counties baseball and hockey leagues

were formed. AG, 17 June 1921, 5; FS, 4 August 1921, 1; SUN, 13 January 1921, 1; 27 January 1921, 6.

18 Other UFO clubs in the area assisted the Rugby club in this initiative. Simcoe County Archives, Rugby Minutes, 9 May 1921.

19 AG, 5 August 1921, 1.

20 Ibid., 11 July 1918, 1.

21 See, for example, Rugby Minutes, 19 December 1921; SUN, 10 July 1918, 7.

22 Goodwyn, *Populist Moment*, xviii.

23 CB, 10 July 1919, 4.

24 LE, 11 September 1918, 8.

25 CB, 4 September 1919, 4; SFRN, 24 August 1920, 5.

26 See, for example, SUN, 16 May 1917, 5; 24 September 1919, 1. See also SFRN, 18 October 1921, 1, 5 and AG, 4 February 1921, 1.

27 CPH, 8 May 1917, 5. On urban perceptions of farmers, see MacGillivray, *Slopes of the Andes*, chapter 3. Distrust towards urban society sometimes took subtle forms. For example, a UFO convention in Lanark County in 1919 started at 1:00 P.M. standard time because the members "were determined to have nothing to do with the troublesome 'new time.'" AG, 3 October 1919, 3. Earlier that year Oro UFO members in Simcoe County had sent a petition to the federal government registering their displeasure at the imposition of Daylight Saving Time. Simcoe County Archives, Oro Minutes, 6 February 1919.

28 SFRN, 8 September 1921, 3.

29 See, for example, the comments of John M. Houldershaw of Simcoe County. SUN, 9 May 1917, 3. Henry John Pettypiece, editor of the *Forest Free Press* and staunch UFO supporter, was critical of the campaign. Ibid., 17 February 1915, 3; 30 August 1916, 3; FS, 12 July 1917, 4. During the war Pettypiece derided Canada's judges for not giving up part of their salary (though they made up to $10,000 annually) while telling others that sacrifices had to be made. SUN, 22 March 1916, 3. On Pettypiece, see Wood, *A History*, 152–3.

30 Of the 10,000 war production questionnaires sent to farmers by the Hearst government in 1917, only 130 were returned. In addition, "mass production rallies elicited virtually no farmer participation ... a provincewide rally at Massey Hall saw farmers represented by the Department of Agriculture officials." Wylie, *Direct Democrat*, 285. On local indifference, see AG, 5 March 1915, 1; Simcoe County Archives, Oro Minutes, 11 December 1917.

31 See Hann, *Farmers Confront Industrialism*, 10–5.

32 OP, 7 February 1918, 8; AG, 28 March 1919, 2; 31 October 1919, 7; CB, 16 October 1919, 5; PC, 3 October 1919, 3.

33 Apparently, the "man and a half per hundred acres" was determined by a Lieut Col Smith, who sat on the London district military tribunal. The rate, however, was an unofficial one. Local tribunals set their own ratios, but it seems that many followed London's lead. SUN, 28 November 1917, 3.

34 Suspicions remained despite assurances to the contrary in the press. In late 1917 the *Forest Standard* reported that Justice Duff's judgement as Central Appeal Judge for Canada "makes it quite clear that farm workers are entitled to exemption on the grounds that production must be maintained ... This judgement ... is binding as law." FS, 13 December 1917, 3; 27 December 1917, 4.

35 The local Food Controller tried unsuccessfully to deliver the petition to Agriculture minister T.A. Crerar personally. Nearly every farmer who spoke on the issue stressed his loyalty but also insisted that farmers be given a "square deal." PC, 23 November 1917, 5, SUN, 21 November 1917, 8. See also PE, 29 November 1917, 3. On page four of the same edition, the names of those who had received temporary exemptions were printed under the headline "Many Exemptions Have Been Granted to Farmers' Sons." Only those who read the article would realize that the exemptions were only temporary. For an account of a similar meeting, see OP, 29 November 1917, 1; SUN, 28 November 1917, 10.

36 CB, 29 November 1917, 7.

37 Cited in Johnston, *E.C. Drury*, 49.

38 SUN, 10 April 1918, 3. Earlier, H.J. Pettypiece had suggested that, if the provincial government were really sincere about the need for everyone to make sacrifices, then it would convert Government House into a treatment centre for wounded returned soldiers. Ibid., 8 November 1916, 3. For his views on wartime taxation see ibid., 22 March 1916, 3.

39 Some newspapers blamed the UFO for the protests. See, for example, PC, 31 May 1918, 5. As seen elsewhere in this chapter, farmers stressed their loyalty at many of these meetings. Although such disclaimers were probably voiced in an attempt to avoid public condemnation, their decision may also have been designed to prevent legal reprisals. In a SUN article ("Farmers Beware! The Police Are on Your Trail," 12 June 1918, 4), it was alleged that a Toronto reporter had informed the police of a meeting in the Toronto Labour Temple in which anti-conscription sentiments were expressed. G. Kennedy of the Toronto Police warned farmers that "from now on no farmer ... will be able to plead ignorance of the Order-in-Council referring to persons spreading disaffection among his Majesty's subjects." It should be stressed that, with the preponderance of such rhetoric, the decision to oppose

conscription was courageous. Even the "ordinarily militant United Farmers of Alberta acquiesced in Ottawa's decision on conscription." Johnston, *E.C. Drury*, 53. In 1918 it was reported that an Ontario farmer was arrested and fined five hundred dollars for stating that "we could not be under any worse Prussian rule than we are now." Anderson et al., *A Political History*, 7.

40 OWT, 9 May 1918, 3. See also OP, 2 May 1918, 1. One Oro farmer took exception to the *Packet's* coverage and claimed that there were many more in attendance than the reported 350. The *Packet's* editor countered by asserting that the report of the meeting was "uncoloured by enthusiasm for or against the object of the meeting." Ibid., 16 May 1918, 6. The *Weekly Times* put the attendance at five hundred, claiming that it was "the largest meeting ever held in Oro Town Hall."

41 BNA, 9 May 1918, 1, 3.

42 SUN, 19 June 1918, 3. Sheehan's reference to cattle destroying crops came from another story he had heard to the effect that a military officer advised a farmer to turn his crops over to his cows if, because of labour shortages, he could not harvest them.

43 A recent example of ignoring farmers' thoughts on WWI can be found in Socknat's *Witness against War*.

44 SUN, 9 January 1918, 3; 14 April 1915, 10. See also ibid., 27 December 1916, 3, for a letter by "Pagan" from Simcoe County.

45 AG, 4 June 1920, 1, 3; OP, 28 October 1920, 5.; OT, 4 November 1920, 7. Even those UFO members who participated in the war effort did not seem to be proud of their actions, choosing to describe them in the mode of "something that had to be done." Compton Jeffs, Progressive candidate in South Simcoe in the 1921 federal election, recalled that he was active in recruiting work (being too old to serve himself), but that he did "not ask for any credit for any action [he] took in the war." Jeffs claimed that he would not have raised the issue at all during the campaign were it not for the fact that former Union candidates were bragging that they had won the war. Jeffs felt that the soldiers, not the politicians, deserved this credit. BNA, 3 November 1921, 2.

46 See, for example, Wood, *A History*, 225–70.

47 SUN, 23 May 1917, 3. Pettypiece also argued that the government used tariffs selectively, noting that urban newspapers had pressured the government into eliminating duties on printing presses and typesetting machines. Ibid., 17 February 1915, 3. See also ibid., 2 November 1921, 6, for his attack on Canada's "cotton lords" who claimed that they could not compete with US firms in the home market without the tariff and yet managed to "sell millions of dollars of goods in the United States and other countries."

48 SUN, 5 March 1921, 3. See also the letter of Charles Stephens of Lambton in the SUN, 17 September 1921, 3.

49 Ibid., 16 April 1919, 8.

50 On Flavelle's saga, see Bliss, *A Canadian Millionaire*, 335–62; Naylor, *The New Democracy*, 82. What is of interest here is the massive public relations campaign that the company undertook to deny the charges. A full-page advertisement from the firm rebutting the allegations appeared in several weeklies. See, for example, *MFP*, 26 July 1917, 3; *SFRR*, 26 July 1917, 4; *CE*, 26 July 1917, 3; *OT*, 26 July 1917, 6; *PC*, 27 July 1917, 7; *AG*, 27 July 1917, 7. The company claimed to have realized a profit of only two-thirds of a cent per pound, but it should be noted that it sold 160 million pounds of meat in the 1917 fiscal year. At two-thirds of a cent per pound of meat sold, Flavelle's firm netted slightly over $1 million in profits. On Gadsby, see Bliss, *A Canadian Millionaire*, 344. Gadsby had particular scorn for businessmen who claimed to be Christians. He attacked Flavelle in 1916 for purchasing a huge quantity of eggs, keeping them in cold storage, and during the winter mixing them with fresh eggs and selling them for a dollar a dozen. "You can figure out for yourself just how precious a few million two-and-a-half-cent eggs would be to Mr Flavelle when sold at the right time." Gadsby then noted that Borden had appointed Flavelle chairman of the Imperial Munitions Board: "From cold storage eggs to shrapnel. From one high explosive to another – both shell games." *SFRR*, 1 March 1917, 4. For other attacks on Flavelle, see ibid., 1 March 1917, 4; 2 October 1917, 4. On the UFO's perspective on business/government collusion, see *SUN*, 8 October 1919, 11.

51 *CE*, 10 July 1919, 1; *CB*, 10 July 1919, 2; *FS*, 3 July 1919, 3.

52 *SUN*, 28 March 1917, 3. See also ibid., 22 September 1915, 3; *OP*, 7 October 1920, 1.

53 *CB*, 17 November 1921, 8. The allegation was formed as a question: "Did not once upon a time the manager of the Sarnia Fence Wire Company publish a statement ... that Col Currie had approached that Company with a view of having the Sarnia Wire Fence Co. join with other fence ... companies to form a trust and ... assured them that a duty would be placed on No. 9, 10, and 12 wire, which would give them control of the fence wire business?" Ibid., 1 December 1921, 4.

54 Ibid.

55 *PC*, 13 July 1917, 8. Kennedy had delivered the same message to Simcoe farmers a few days earlier. At that meeting E.C. Drury noted that the press "fashioned public opinion" about farm conditions without knowing anything about the subject. *OP*, 12 July 1917, 6. Good (*Farmer Citizen*, 98) maintained that the press was chiefly responsible for the "corruption of social taste."

56 Pettypiece's *Forest Free Press* had, he boasted, a circulation of 1,600 in 1920. AO, RG 3, Series 4, Box 22, File 03–04–0–105, Pettypiece to Drury, 10 May 1920.

57 In 1916, for instance, he drafted an article for the *Toronto Globe* which pointed out that, in addition to benefiting from the tariff, Canadian manufacturers also received over $3.25 million in customs refunds and drawbacks in 1915 alone. The *Globe's* editors refused to publish the piece, contending that "there is no drawback allowed when manufactured for home consumption." Pettypiece maintained that he could prove that such refunds had been paid since 1907. For Pettypiece, the incident served to show how the Big Interests prevented the truth from reaching the public. *SUN*, 30 August 1916, 3, also 28 March 1917, 3. *Almonte Gazette* editor James Muir sometimes criticized the press. He particularly abhorred the "voluntary censorship" newspapers conducted during the War. *AG*, 2 August 1918, 4.

58 See, for example, ibid., 21 March 1919, 4.

59 East Simcoe UFO MPP J.B. Johnston was a critic of the press, which he saw as a tool of the old political parties. *OT*, 16 October 1919, 6; *OP*, 17 June 1920, 1; *SUN*, 1 October 1919, 1, 4.

60 *OP*, 1 December 1921, 6; *OT*, 1 December 1921, 3.

61 Moreover, their responses reveal how highly literate some farmers were. A study undertaken in 1915 surveyed four hundred farmers in four Ontario counties. Among its findings were that sixty-seven percent of those surveyed took an agricultural paper, seventeen percent received story magazines, and seventy-seven percent subscribed to a daily newspaper. *OSA*, 17 February 1916, 4. See also MacGillivray, *The Mind of Ontario*, 52.

62 *SUN*, 4 December 1918, 3. "Farmer" had approached the *Orillia Packet* (1 May 1919, 8) to try to correct "an amazing amount of nonsense" that a *Packet* reporter had written about rural depopulation. Several examples of such sentiments can be found. Alfred G. Tate of Haliburton County had written of the relationship between the government and the press during the war: "As to the real dangers and necessities incurred by the war, let the Government come out flat-footed and let the people know what is required and why. But when the *Toronto Star* says … 'We must have faith that what the London, England, authorities do is for the best,' you will have to pardon us hayseeds, but we have not been used to running countries on faith. We want facts." *SUN*, 6 March 1918, 3.

63 Some newspapers distorted incidents to serve their own ends. In 1920 the *Smiths Falls Record-News* reported a speech by Drury in which he allegedly stated that the UFO/ILP coalition could not continue. Drury had, in fact, alluded to "Broadening Out," and buried in the article it

was noted that he had called upon all sectors in society to form a "People's Progressive party." The headline, however, "UFO and Labor to be Divorced," conveyed a decidedly different impression. *SFRN*, 7 December 1920, 1. If newspapers did not distort, they certainly often intended to embarrass. See *CB*, 12 May 1921, 2, in which it was reported that UFCC president R.W.E. Burnaby claimed he had been offered a thousand-dollar bribe to use his influence to secure an appointment for an individual as a government purchasing agent. Burnaby was asked to reveal the person's name to the Standing Committee on Privileges and Elections, but he refused to do so. The *Bulletin* then listed an "Honour Roll" of UFO/ILP MPPs who had defeated a Liberal motion in the Legislature to compel Burnaby to divulge the name of the individual. Some local papers, however, did try to balance critiques of the UFO with accounts of how other newspapers deliberately distorted the farmers' position. See, for example, *FS*, 2 February 1921, 8.

64 *OP*, 11 November 1920, 10. Some newspapers gave as good as they got. See *CE*, 10 July 1919, 4, which contained jabs at Drury and Morrison in two separate articles.

65 *CB*, 24 November 1921, 5. Many newspapers actively opposed the UFO. The *Barrie Northern Advance* was a particularly harsh critic of the movement, and it often reprinted attacks on the UFO from other newspapers. See, for example, *BNA*, 17 November 1921, 5.

66 Baker also alleged that provincial officials had prepared over thirty briefs to prosecute combines in Ontario charged with restraint of trade, but that "our Crown Attorney has so far refused to prosecute these charges." *CB*, 16 October 1919, 5. See also the comments of Compton Jeffs of Simcoe (*SUN*, 1 October 1919, 1, 4), South Lanark candidate W.I. Johnson (*SFRN*, 30 September 1919, 1; *PE*, 2 October 1919, 3), and North Lanark candidate Hiram McCreary (*PC*, 3 October 1919, 4).

67 *SUN*, 11 December 1918, 3. Earlier, Taylor had written that "Manitoulin has shown the way," and that every riding should nominate UFO candidates. If they were elected, it would "end once and for all Government by the rich ... Hearst seems to fear that the Manitoulin man will be lonesome in the House. Let us then see to it that we ... give him lots of company" at the next election. Ibid., 6 November 1918, 3.

68 *OT*, 17 November 1921, 4; *BNA*, 17 November 1921, 1, 5; *OT*, 24 November 1921, 4. See also the comments of Compton Jeffs, South Simcoe candidate, *BNA*, 8 September 1921, 6; and 3 November 1921, 3, where he argued that the railroads had not been built to aid the people who relied upon them. Instead, they were built "for the money that could be got out of them."

69 OT, 24 November 1921, 4.

70 CB, 25 September 1919, 1, 4; FS, 3 Jul 1919, 3; SUN, 1 October 1919, 1, 4.

71 AG, 19 September 1919, 1; SFRN, 16 September 1919, 1; CPH, 17 September 1919, 1. See also the comments of W.I. Johnson. PC, 26 September 1919, 8; 3 October 1919, 1; SFRN, 30 September 1919, 4; PE, 2 October 1919, 3; AG, 3 October 1919, 3.

72 Allen, *Social Passion*, 264.

73 OP, 16 October 1919, 2; OT, 16 October 1919, 6; OWT, 16 October 1919, 5. See also SUN, 1 October 1919, 1; CB, 16 October 1919, 5.

74 In North Lanark, UFO candidate Hiram McCreary was referred to by local members as the riding's only temperance candidate. See the comments of local farmer W.H. Robertson in AG, 17 October 1919, 12. See also ibid., 10 October 1919, 6. According to the *Gazette*, McCreary's upset victory was attributable, at least in part, to his stand on prohibition. Ibid., 24 October 1919, 1–2. On the older parties, see the comments of Lambton East Tory candidate J.B. Martyn (FS, 16 October 1919, 4). The commitment of old party candidates, however, was a matter of some suspicion among temperance adherents. As MacGillivray notes: "A fact always forgotten by those readiest to sneer at the Ontario temperance movement, was that it operated from the beginning to end against powerful economic forces. The brewers, distillers, and tavern keepers, and all who were financially interested in their welfare, understood and hated it. The politicians ... knew how helpful tavern-keepers were in organizing support at elections. Not surprisingly, the politicians often proved unreliable, if not treacherous, supporters of temperance." MacGillivray, *Mind of Ontario*, 56.

75 AG, 2 December 1921, 2.

76 LE, 11 September 1918, 8; SUN, 23 July 1919, 10; Simcoe County Archives, Oro Minutes, 6 November 1919.

77 AG, 11 November 1921, 2.

78 AG, 4 June 1920, 1, 3. That some UFO leaders staunchly supported the Ontario Temperance Act certainly helped the farmers' cause. E.C. Drury was renowned for his personal pro-temperance stance. Indeed, he spoke on the issue as often as he could. See, for example, OT, 14 April 1921, 3.

79 Webster was born in Simcoe County in 1868 and attended local schools and the Ontario Agricultural College. He began farming near Creemore in 1890 and in 1902 married Alice Hollingsworth, a member of the Women's Institute and a founding member of the United Farm Women of Ontario. Webster was prominent in several early agrarian organizations, such as the POI, FI, and FA. He helped establish an apple-growers' cooperative, assisted in forming a local telephone

company, and was a founding member of the Madill's Corners UFO Club. *SUN*, 29 September 1927, 9.

80 Ibid., 11 December 1918, 3. A.J. Forsyth of Barrie had earlier expressed similar sentiments but for different reasons. He favoured the single tax because it would rid communities of land speculators. Ibid., 10 October 1917, 3.

81 Ibid., 2 April 1919, 8. A year earlier the *SUN* quoted from Wilhelm von Humboldt's *The Sphere and Duties of Government*, in which it was argued that, as the state passes laws to make itself relevant, there arises a need to increase state revenue and the number of officials. Public affairs becomes more complex to the point that it requires "an incredible number of persons to devote their time to its supervision, in order that it may not fall into utter confusion." People who would otherwise engage in creative and useful tasks become laden with bureaucratic duties. Thus, people tend to move away from self-reliance. "Hence it arises that every decennial period the number of public officials increases, while the liberty of the subject proportionately declines." Ibid., 25 September 1918, 3. On Humboldt's connection to anarchism, see Chomsky, *Chomsky Reader*, 147–51; Brown, "Anarchism, Feminism, Liberalism and Individualism," 27.

82 *SUN*, 4 September 1918, 3; 14 August 1918, 3.

83 Ibid., 1 October 1919, 1.

84 *SFRN*, 30 September 1919, 1; *PE*, 2 October 1919, 3; *PC*, 17 October 1919, 8. See also *SFRN*, 24 August 1920, 5; *AG*, 17 September 1920, 1. North Lanark UFO candidate Hiram McCreary assured voters that he could not be "swayed by the Big Interests" and that he was the "candidate of the people." Ibid., 10 October 1919, 6.

85 *SUN*, 13 February 1918, 3. Heick's *Propensity to Protect* fails to convey the vehemence with which the debate was conducted at the local level. For example, the Oro Station club sent two delegates to the annual Board of Agriculture meeting in Orillia in 1917. Both delegates voted for the legalization of margarine and were subsequently "severely criticized for using their votes in such a manner … but for the fact that the club had appointed its own delegates and to abide by their influences no penalty could be imposed." Simcoe County Archives, Oro Minutes, 5 July 1917.

86 *SUN*, 11 December 1918, 3. Ibid., 13 August 1919, 11. At one stage during the 1919 campaign, a UFO member from North Lanark suggested that Hiram McCreary be referred to as "the candidate of the farmers of the North Riding." Although many members agreed with the suggestion, it seems that it was not acted upon. *SFRN*, 16 September 1919, 1; *CPH*, 17 September 1919, 1; *AG*, 19 September 1919, 1.

87 Good, *Farmer Citizen*, 117; LCA, East Lambton Minutes, 24 July 1919.

88 *CB*, 3 April 1919, 3. The decision to contest elections came from the local clubs, not the central office. In fact, the UFO leaders spent most of the 1919 election campaign catching up on what was going on in the localities. See Drury, *Farmer Premier*, 81–4; Good, *Farmer Citizen*, 120. The desire for political action was not restricted to provincial politics in 1919. For instance, some Lambton farmers advocated uniting politically at the municipal level, so as to defeat an "autocratic" county council that had passed roads legislation that favoured urbanites. *SUN*, 5 March 1919, 3.

89 The decision to field a candidate or not was left to the clubs as well. When T.P. Loblaw of the central UFO cancelled his scheduled talk at a UFO meeting in Smiths Falls in August 1919, one member suggested that the fielding of a UFO candidate be discussed. However, the nomination of a candidate appears to have been taken for granted and, as a result, members discussed instead what qualities the candidate should possess. *SFRR*, 26 August 1919, 4; *AG*, 29 August 1919, 6. The possibility of fielding an UFO candidate in Lanark was alluded to prior to this meeting. See ibid., 21 March 1919, 4; *SFRR*, 20 May 1919, 4; *SUN*, 11 June 1919, 10.

90 In Centre Simcoe, three candidates were nominated for the 1919 provincial election. Although James Martin received more than fifty percent of delegates' votes on the first ballot, the second place candidate insisted on a second ballot before resigning, because it "would be a cowardly thing ... if I am not fit to represent you, put me out." *CE*, 11 September 1919, 7. The delegates agreed to alter the voting procedure and stage a second ballot, even though it was not necessary.

91 LCA, East Lambton Minutes, 24 July 1919. This structure was also in place for the 1921 contest. *PAT*, 13 October 1921, 1. In West Lambton in 1920, it was decided to field a joint UFO/ILP candidate in the next federal election. The convention to select a candidate was structured much like East Lambton's. *SUN*, 26 June 1920, 9.

92 See, for example, LCA East Lambton Minutes, 24 September 1919, where it is noted that members felt that the campaign would cost $510. Each township director was given a target, ranging from $80 for Euphemia to $110 each for Brooke, Warwick, and Plympton. Evidently, that figure was attained. West Lambton raised $311 for the 1919 campaign, with no member contributing more than one dollar. *SUN*, 3 December 1921, 9. In a convention to determine whether or not to field candidates in Lanark in 1919, delegates from North Lanark voted to pay all expenses incurred by the candidate, whereas South Lanark delegates refused to do so, arguing that they did not have the authority to bind their clubs to such an agreement. *SFRN*, 9 September 1919, 4; *AG*, 12 September 1919, 1; *PC*, 12 September 1919, 2. Simcoe

Clubs also raised money for UFO candidates. Simcoe County Archives, Edenvale Minutes, 8 October 1919.

93 The recall was a contentious issue in some ridings. In Centre Simcoe in 1919 there was a heated debate on the subject during a nomination meeting. One nominee, J.T. Simpson, opposed the measure, arguing that if people did not have confidence in the candidate to use his judgment, then he did not wish to stand as a nominee. *CB*, 4 September 1919, 4. East Lambton UFO members insisted that any candidate they nominated sign recall papers. Simcoe County Archives, East Lambton Minutes, 20 August 1919.

94 Quoted in Wood, *A History*, 278.

95 *CSN*, 16 August 1919, 3; *SUN*, 20 August 1919, 1. *SFRN*, 16 September 1919, 1; *CPH*, 17 September 1919, 1; *AG*, 19 September 1919, 1. See also Halbert's comments in *CE*, 18 September 1919, 2; and A.A. Powers's remarks to Lanark farmers in *AG*, 21 March 1919, 4; *PC*, 21 March 1919, 1.

96 *CB*, 2 October 1919, 4; Simcoe County Archives, East Lambton Minutes, 20 August 1919; *SUN*, 1 October 1919, 1, 4.

97 *SFRN*, 20 March 1919, 4; *AG*, 17 October 1919, 12; *PC*, 10 October 1919, 4.

98 *FS*, 23 June 1920, 1; *OT*, 17 November 1921, 4, 24 November 1921, 6; *OP*, 24 November 1921, 1.

99 See, for example, the editorial regarding the annual Smiths Falls UFO meeting in the *SFRN*, 2 December 1920, 6.

100 *AG*, 21 October 1921, 1, 4. See also *PE*, 20 October 1921, 4; *PC*, 21 October 1921, 2.

101 *AG*, 2 December 1921, 2. *AG*, 21 October 1921, 2. Anderson, was also vice-president of the Lanark Mutual Fire Insurance Company and considered himself a lifetime "student of Political Economy." *SFRN*, 1 December 1921, 4.

102 *SFRN*, 6 December 1921, 5. The committee's platform, which was adopted by the Lanark County Progressives, is found in Appendix Q.

103 See Appendix R for the results of federal and provincial elections involving UFO/Progressive candidates in Lambton, Simcoe, and Lanark.

104 *OP*, 27 October 1921, 1. The remark drew outraged howls from the press, especially since the UFWO meeting was to have been a social gathering. UFO member Lawrence Cooper defended the two men by claiming that they had been invited by the UFWO to discuss how the tariff affected women. Ibid., 3 November 1921, 8.

105 *CB*, 17 November 1921, 8, 24 November 1921, 2.

106 Ibid., 1 December 1921, 4.

107 *BNA*, 17 November 1921, 1; *OT*, 24 November 1921, 4. As did the Lanark Progressives, Swindle favoured a direct tax on land values;

graduated taxes on income, inheritance, and corporate profits; direct legislation; proportional representation; Senate reform; an eight-hour day; public control of the railways, public utilities, and natural resources; disability insurance; old age pensions; unemployment insurance; and "total disarmament." The only significant difference in Swindle's campaign was that he called for "the exclusion of all Asiatics." Ibid., 1 December 1921, 3; *OP*, 1 December 1921, 6. Given that virtually all UFO clubs argued for racial harmony among groups in Canada, it is difficult to understand why Swindle advocated this measure, other than to appeal to the nativist tendencies in some urban workers. On this theme, see Naylor, *The New Democracy*, 103.

108 *OT*, 13 October 1921, 6, 24 November 1921, 1; *CB*, 24 November 1921, 4. See also the comments of South Simcoe Progressive candidate Compton Jeffs (*BNA*, 8 September 1921, 6; 3 November 1921, 2), and those of West Lambton candidate R.J. White (*FFP*, 16 November 1921, 3). On Ross, see *OT*, 13 October 1921, 6; Johnson, *The Canadian Directory*, 510.

109 See, for example, W.L. Morton, *The Progressive Party*, 118.

110 As will be seen, returned soldiers were often represented separately at UFO/ILP conventions. The evidence suggests, however, that they were not a particularly vocal group. Although this should not be construed to mean that returned soldiers as a group were unimportant, the focus here is on the more public deliberations among ILP and UFO members. On returned soldiers, see Morton and Wright, *Winning the Second Battle*, 162: Even they acknowledge that, regarding the 1921 campaign, "Considering the veterans' role in 1917 and the anathema the GWVA [Great War Veterans' Association] had pronounced on the Unionists in 1920, veterans had remarkably little to say in the [1921 campaign]."

111 *CE*, 11 September 1919, 8; *CB*, 11 September 1919, 1;. *CSN*, 18 October 1919, 1. In West Lambton, the UFO and ILP amalgamated in 1920 to fight the next federal election. *PAT*, 22 July 1920, 8.

112 *AG*, 19 September 1919, 1.

113 Simcoe County Archives, East Lambton Minutes, 24 July 1919, 20 August 1919. That East Lambton was largely rural probably explains the decision not to affiliate formally with the ILP.

114 *SFRN*, 2 October 1919, 1; 16 October 1919, 1; *AG*, 10 October 1919, 11. In West Lambton, UFO members nominated Jonah M. Webster, while the ILP nominated James S. Crawford. *SUN*, 13 August 1919, 11.

115 *OP*, 28 August 1919, 1, 8. Although not reported in the local press, a *Toronto Star* article claimed that the meeting was an acrimonious one, with R.H. Halbert allegedly saying to Foster, "It looks to me as if Labor is afraid of the farmers," to which Foster was reported to have replied, "Labor is afraid of no one. There is a chance of a cleavage

here, and I warn you to avoid boasting." None of this was reported in local press stories, but the *Star* article was reprinted in CB, 4 September 1919, 6. In addition, a week after the original convention, the *Packet* quoted an unnamed local farmer who insisted that farmers wanted their own candidate, and that they did not wish to be seen as connected with the ILP. *OP*, 28 August 1919, 8. One wonders if the *Packet*, not known for its sympathy for farmers or for labour, distorted the farmer's statement or even fabricated it.

116 Ibid., 11 September 1919, 1, 6. Local papers made much of the irony of farmers, who complained about not having enough of their own at Queen's Park, nominating an industrialist's agent as their candidate. As the *Packet* emphasized, Johnston worked neither on a farm nor in a factory, and he had not served in the army. Ibid., 11 September 1919, 9; 9 October 1919, 1; *OT*, 25 September 1919, 4.

117 On Johnston's shortcomings (mainly his evasiveness and his lack of elocution skills), see ibid., 16 October 1919, 6; *OWT*, 16 October 1919, 6; *OP*, 16 October 1919, 8.

118 At an Oro Station UFO meeting held shortly after the election, Johnston and Foster both spoke on the differences between the two groups, apparently to underscore how difficult the next few years would be. Simcoe County Archives, Oro Minutes, 23 October 1919, 1. Tensions between these two groups flared occasionally after 1919. In Lanark, the ILP chose Duncan H. Gemmell in 1920 to contest the next federal election and asked the Smiths Falls UFO to endorse the choice. Although some farmers believed that the two groups should amalgamate, others felt that the ILP's call for an eight-hour day made an alliance impossible. UFO members ultimately decided that it "would not endorse anything but an open convention" to select a candidate. One farmer, James Porter, said that Almonte Labourites had recently stated that they would field a candidate regardless of what the farmers decided to do: "We offered to go before them in open convention ... but they didn't want that." By that time, the ILP claimed a membership of four thousand in the county, a claim that some UFO members dismissed as an exaggeration. *SFRN*, 3 November 1920, 2.

119 *OP*, 7 October 1920, 1; *OT*, 7 October 1920, 1, 4.

120 Chew had contested the 1908, 1911, and 1917 elections and had been elected in 1911.

121 *OP*, 6 October 1921, 1.

122 According to one account, the UFO delegates were willing to abandon the agreement on the condition that the nominee be either a farmer, a worker, or a returned soldier. "Against a manufacturer they were adamant." *CE*, 13 October 1921, 2.

123 The account of this meeting is a composite derived from several area newspapers: ibid.; *OT*, 6 October 1921, 1, 4; *CE*, 13 October 1921, 2. Manley Chew was later selected by the Liberals to contest the riding, and he received the support of the *Orillia Times*, which castigated the farmers for selfishly keeping their own candidate in the contest, thus opening up the possibility of splitting the non-Tory vote. The editor could not see why the UFO would allow this to happen, considering its object – "the defeat of the Meighen government." Undoubtedly this must have come as a surprise to some Progressives, who believed that the object was to defeat the party system, not just one party. *OT*, 20 October 1921, 4; *OP*, 13 October 1921, 1; 20 October 1921, 1.

124 See Staples, *The Challenge*, 156.

CHAPTER FOUR

1 Thompson with Seager, *Canada 1921–39*, 27–8; Ontario, Chief Election Officer, *Electoral History of Ontario*, J5-J6; Beck, *Pendulum of Power*, 160–1.

2 The political success of the UFO and the Ontario Progressives in 1919 and 1921 was, in fact, a statistical aberration. Although in 1919 the UFO won more seats than any other party, it received only 21.7 percent of the popular vote. The Tories and Liberals received 34.9 percent and 26.9 percent respectively, but they elected fewer MPPs. In 1921, as mentioned above, Ontario Progressives received 27.7 percent of the popular vote. Thus, although there was momentum to that point, the chances of the UFO repeating its success in 1923 was slim, especially given the overall inexorable shift in the rural/urban population figures. The chances of further political success would be limited if the movement were to remain strictly rural. Ontario, Chief Election Officer, *Electoral History*, J5-J6; Beck, *Pendulum of Power*, 160–1.

3 The emphasis on the UFO's leaders is evident in the *SUN*, but other pro-UFO newspapers focused on these individuals to an even greater extent. James Muir, editor of the *Almonte Gazette*, is a case in point. See, for example, *AG*, 4 June 1920, 6; 30 April 1920, 1; 14 January 1921, 3; 23 July 1920, 7; 21 January 1921, 1; 5 August 1921, 1.

4 On the early years of the UFO government, see Johnston, *E.C. Drury*, 68–82.

5 The *Almonte Gazette* complained that Drury and Doherty remained without seats because many UFO clubs "refused to allow their elected representatives to resign and thus make the way for non-elected ministers. *This failure to provide seats for ministers shows a serious lack of party loyalty.*" *AG*, 9 January 1920, 3 (emphasis added). For a newspaper that vilified blind obedience in the old parties, this was a curious stance to

adopt. On the issue, see also *FS*, 25 December 1919, 1; *AG*, 23 January 1920, 2; Bristow, "Agrarian Interest," 35.

6 For example, in October 1920 farmers learned of Drury's reluctance to oppose proposed Bell telephone rate increases on the ground that the federal government owned "more than one-half the railway mileage of Canada and the Provincial Government did not want to clash with the Federal power." At that time Bell had an exclusive contract with Canada's railways regarding its long-distance lines and was regulated by the federal Board of Railway Commissioners. Armstrong and Nelles, *Monopoly's Moment*, 163–86. On the 1920 rate increase, see ibid., 276–7. Drury later stated his opposition to the proposal. *FS*, 7 October 1920, 1.

7 Manning Doherty expressed similar sentiments at about the same time. *FS*, 23 June 1920, 1. See also J.J. Morrison's comments, made in Lambton a short time later, in which he noted that the time was ripe to secure "some new means for the obtaining of the concensus [*sic*] of opinion of the people on which, and only on which, the country should be ruled." Ibid., 1 July 1920, 3, also 8 July 1920, 1.

8 On UFO legislation, see Johnston, *E.C. Drury*, 149–65.

9 To be fair, many of the planks fell under federal jurisdiction, but there were some that could have been acted upon by the provincial government but were not. The province, for instance, would have been within its rights to implement programs for returned soldiers; to extend cooperative agencies; and to implement electoral reforms. For the Platform text, see Appendix S.

10 *SFRN*, 31 May 1923, 3; *PC*, 1 June 1923, 4; *AG*, 1 June 1923, 4. To be fair to Drury, UFO/ILP MPPs only added up to fifty-five in a legislature of 111. Drury was understandably careful because he had to count on the support of other MPPs to stay in power. Johnston, *E.C. Drury*, 81–2.

11 *OP*, 24 May 1923, 2–3; *BNA*, 24 May 1923, 5; *CE*, 31 May 1923, 4. For accounts of other speeches Drury made in Simcoe during the campaign, see *OT*, 14 June 1923, 4; *OP*, 14 June 1923, 2. Commenting on a rally held in Orillia, local newspapers noted that there was none of the enthusiasm that had characterized UFO gatherings in the past. The hall in which the rally took place was not even filled. *OT*, 14 June 1923, 4; *OP*, 14 June 1923, 2. Drury had defended his stand on race-track betting a year earlier in Oro Station. *BNA*, 27 July 1922, 5.

12 The transferable vote is a system in which a voter lists candidates in order of preference. The candidate with the fewest first-place votes is dropped, and his or her second-place votes go to the remaining candidates. This continues until one candidate emerges with at least fifty percent plus one. Under proportional representation, seats in the legislature are awarded parties according to the percentage of votes they are able to obtain.

13 *OP*, 24 May 1923, 2–3; *BNA*, 24 May 1923, 5; *CE*, 31 May 1923, 4. See also *AG*, 27 April 1923, 2
14 Ibid., 10 Augst 1923, 1.
15 *BE*, 1 October 1925, 1–2; *CB*, 1 October 1925, 1; *CE*, 1 October 1925, 1, 5. Ross seems to have been a competent MP, one who tried to adhere to the UFO principles. During the 1923 provincial campaign, he claimed that there were only two political parties: one that represented the moneyed interests (the Liberals and Conservatives), and one that worked on behalf of the "common people." *BNA*, 24 May 1923, 5.
16 *BE*, 1 October 1925, 1–2; *CB*, 1 October 1925, 1; *CE*, 1 October 1925, 1, 5.
17 There were, however, a few exceptions. In a debate on the topic with his Conservative opponent, Drury argued that the tariff was iniquitous and inflationary, and that it fostered combines that allowed manufacturers to reduce production and thereby raise prices. *BE*, 29 October 1925, 1, 13.
18 *AG*, 4 June 1920, 1, 3. On Halbert's Orange membership, see Pennefather, "The Orange Order," 173.
19 *OP*, 16 November 1922, 1; *OT*, 16 November 1922, 1.
20 *AG*, 2 June 1923, 1.
21 Ibid., 8 June 1923, 2.
22 On the pre-war growth of the provincial bureaucracy, see Hodgetts, *From Arm's Length to Hands On*, 110–39.
23 *AG*, 17 June 1921, 1; *OT*, 4 November 1920, 7; *SUN*, 17 December 1919, 10.
24 *AG*, 25 May 1923, 1. During the 1923 campaign South Simcoe UFO MPP E.J. Evans felt compelled to explain why North Simcoe had received more money for roads than South Simcoe had. *SUN*, 7 June 1923, 3.
25 *BNA*, 24 May 1923, 5; 27 July 1922, 5; *OP*, 21 June 1923, 1, 12; *OT*, 21 June 1923, 5. *CE*, 1 October 1925, 4.
26 *AG*, 23 April 1920, 1; 30 April 1920, 1; Hallowell, *Prohibition in Ontario*, 76–7; Johnston, *E.C. Drury*, 157–9. McCreary's election advertisements were less extensive during the 1923 campaign than they had been in 1919. Whether this was due to a lack of funds or to arrogance is unknown. Only late in the campaign, when it dawned on his supporters that he might not be re-elected, did advertisements begin to appear.
27 *AG*, 22 June 1923, 4.
28 Ibid., 9 October 1925, 1; 16 October 1925, 4. Gemmell also sat on the Smiths Falls Town Council. By waiting until mid-October to announce his candidacy, he complicated the political situation in Lanark. The Liberals, not knowing his plans by early October, nominated Dr E.H. Wickware as their candidate. Until that time it was understood among Lanark Liberals that if Gemmell decided to run, they would not run a

candidate against him. Wickware later declined the nomination. Ibid., 2 October 1925, 1; PC, 9 October 1925, 5; 16 October 1925, 6.

29 AG, 23 October 1925, 5; CPH, 21 October 1925, 8; PC, 23 October 1925, 8.

30 Ibid., 16 October 1925, 6; AG, 16 October 1925, 5.

31 Ibid., 23 October 1925, 5; PC, 23 October 1925, 8.

32 Gemmell had a chequered history of contesting elections in Lanark. Nominated by the Progressives in 1922 to run in federal a by-election, with three weeks remaining in the campaign he declined to accept the nomination, citing time and logistical constraints. Owing to the timing of his decision, the Progressives did not field a candidate. SFRN, 14 November 1922, 1, 5; AG, 17 November 1922, 1; PC, 17 November 1922, 2. He ran again in the 1926 provincial election as a Prohibition Unionist and was endorsed by the South Lanark Progressive Association. Much to its chagrin, he withdrew only days before the election. SFRN, 11 November 1926, 7; 18 November 1926, 1; PC, 12 November 1926, 3; AG, 12 November 1926, 1.

33 FS, 21 June 1923, 1.

34 FFP, 24 December 1925, 8; FS, 24 December 1925, 4.

35 Ibid., 25 November 1926, 2,6.

36 In early 1925, when W.E. Raney succeeded Manning Doherty as leader of the UFO at Queen's Park, Oke stated that he would not accept Raney as leader, and that he would henceforth sit as a "clear" UFO MPP. It was no secret that Oke opposed Raney's leadership, if for no other reason than he was not a farmer. Part of his statement read: "It is my understanding that I was elected to the Legislature to uphold the rights of the farmer in legislative affairs, in accordance with the principles of the UFO organization. This being my conception of the circumstances under which I was elected, I cannot … recognize any … allegiance to a group in the Legislature which … is in no sense charged with the advancement of UFO principles and ideals. This being the case, I emphatically deny that Mr Raney is the leader of a group which is in any way entitled to be called the UFO." Ibid., 29 January 1925, 1.

37 Goodwyn, *Populist Moment*, xvii.

38 On possessive individualism, see Macpherson, *Political Theory*, 263–4. Macpherson contends that possessive individualism is the dominant societal paradigm in capitalist society, a view I share.

39 The picnics and other social events that continued to characterize the movement enabled farmers to exchange ideas and helped to alleviate feelings of isolation, thereby fostering agrarian solidarity. It should be mentioned that these events were often staged in order to raise money or goods for fire or famine victims, and other charities. However, they featured speakers from the central UFO and from the movement's legislators, so that the local members were drawn into the battles of the

central office. See, for example, *SUN*, 22 June 1922, 4; *OP*, 9 November 1922, 8; 29 March 1923, 3; *FS*, 24 January 1924, 8; 21 February 1924, 2; *SFRN*, 28 March 1922, 1; *AG*, 7 March 1924, 1; 27 June 1924, 1; 5 March 1926, 1. As time went by, even UFO social events were not free from commercial influences. The Ramsay UFO picnic in 1925, for instance, featured Ford officials, who presented motion pictures on tractor operations and other "educational" subjects to an apparently appreciative audience. *SUN*, 25 June 1925, 8. Prominent agrarian politicians visited local clubs fairly frequently. During 1920–21, for example, Lanark was visited at least two times each by R.H. Halbert and J.J. Morrison, and also by Senator W.L. Church of North Dakota, A.A. Powers of the United Farmers Cooperative Company, R.H. Grant, Drury's minister of Education, and Battleford MP H.O. Wright. *SFRN*, 8 July 1920, 7; 17 February 1921, 4; *AG*, 23 July 1920, 6; 18 March 1921, 8; 17 June 1921, 1.

40 *SUN*, 26 November 1919, 10; 3 December 1919, 10.

41 Simcoe County Archives, Oro Minutes, February 1922. There are also indications that the UFO provided an example to other citizens. Soon after the 1919 provincial election the Fifty Sunday Meeting Association was established at Smiths Falls. Its object was to encourage people to "take a greater interest in the town's affairs." Whether this group patterned itself after the UFO is uncertain, but the fact that some prominent UFO members were involved in its creation is significant. *SFRN*, 30 December 1919, 7.

42 *SUN*, 26 June 1920, 9.

43 *OP*, 4 May 1922, 4; 18 May 1922, 3; 6 July 1922, 6; *CSN*, 27 October 1923, 1–2. In 1923 East Simcoe UFO supporters expected their candidate not to back any coalition government with any other political party except ILP or Soldier members, without first calling a convention "and abiding by [its] decision." *OP*, 24 May 1923, 2–3; *BNA*, 24 May 1923; *CE*, 3 May 1923, 4.

44 LCA, East Lambton Minutes, October 1922; *FS*, 27 January 1921, 1; *SUN*, 22 June 1922, 4.

45 LCA, East Lambton Minutes, 14 April 1923; *SCO*, 23 April 1923, 1.

46 LCA, East Lambton Minutes, 5 May 1923. Their specific concerns were not recorded in the minutes. On the Adolescent School Attendance Act, see Johnston, *E.C. Drury*, 90, 190. Oke was also one of seven UFO MPPs who opposed Drury's Civil Service Superannuation Act. Ibid., 128. At roughly the same time the executive of the South Lanark UFO Riding Association passed a similar resolution of support for Drury's administration. *SFRN*, 10 May 1923, 1.

47 LCA, East Lambton Minutes, 10 May 1923. On the civil service superannuation legislation, see Johnston, *E.C. Drury*, 128–9. On the removal of property requirements, see *OT*, 25 November 1920, 4.

48 LCA, East Lambton Minutes, 15 May 1923; *SCO*, 16 May 1923, 1–2.
49 Ibid., 21 June 1923.
50 The UFO elected seventeen MPPs in the 1923 election, three more than the third-place Liberals, but refused to become the Official Opposition.
51 LCA, East Lambton Minutes, 21 July 1923.
52 Ibid., 9 November 1923.
53 *SUN*, 14 December 1922, 5.
54 *FS*, 19 April 1923, 4; 7 June 1923, 4.
55 *SUN*, 22 August 1922, 5.
56 Ibid., 29 August 1922, 5.
57 Ibid., 16 July 1921, 3; *SFRN*, 14 June 1923, 1, 4; *PC*, 22 June 1923, 3.
58 *SUN*, 8 October 1925, 8.
59 At the December 1925 UFO convention, delegates voted to sanction political action once again. *AG*, 18 December 1925, 3; 25 December 1925, 1.
60 *SUN*, 4 March 1926, 8. During the 1926 federal election campaign Webster observed that three distilleries had contributed a total of $274,737 to the Conservative and Liberal parties, and he wondered if these contributions had anything to do with the Customs Scandal. There had been no criminal investigation, but if the UFO had done the same thing, then "a tremendous howl would go over the land. Why did not Hon. H.H. Stevens move in the committee for prosecution of these companies, when he knew the facts? The answer is only too well known." Ibid., 2 September 1926, 14.
61 *FFP*, 24 December 1925, 8; *FS*, 24 December 1925, 4.
62 Ibid., 24 December 1925, 4; *FFP*, 24 December 1925, 8.
63 *PAT*, 24 September 1925, 1. Political musings continued well after the UFO had declined as a political force. See the comments of J.E. Capes of Lambton in *SUN*, 7 April 1927, 4, and the comments of H.J. Pettypiece in *FFP*, 29 October 1925, 2.
64 "Senex" of Orillia noted in 1923 that the UFO had quashed several ILP initiatives, including a bill providing for the distribution of free textbooks to all public schools and another establishing proportional representation in municipal elections. Senex was particularly bemused at the latter action, especially since PR was a plank in the Farmers' Platform (see Appendix S). *OT*, 8 March 1923, 7. See also *AG*, 3 February 1922, 7.
65 *PE*, 24 May 1923, 2; *PC*, 25 May 1923,4; *AG*, 10 December 1920, 1. See also *SFRN*, 14 June 1923, 1, 4; *PC*, 22 June 1923, 3, 8.
66 *OP*, 10 May 1923, 1, also 17 May 1923, 1.
67 *OT*, 31 May 1923, 1. At a Liberal convention in Coldwater held at roughly the same time, local Liberal Association president Dr J.A. Harvie said that the Liberals had a good chance of taking the seat if

an alliance with labour was arranged. When it became clear that labour would not assist in selecting a candidate, the meeting turned into a UFO-bashing affair. Ibid., 31 May 1923, 5; *OP*, 31 May 1923, 2.

68 Ibid., 31 May 1923, 1. No local newspaper carried the story until after it was reported in the *Globe*, and the allegations appear not to have been followed up.

69 Palmer, *Capitalism Comes to the Backcountry*, 17–8.

70 Occasionally, farmers learned of the efforts of the old parties to defeat their movement at the polls. In 1920 the UFO won a by-election in East Elgin despite "the titanic efforts made by the Conservative forces to win at any price ... There was not only apparently money in unlimited quantities for election expenses, but an array of government speakers." *PC*, 26 November 1920, 2. Generally, candidates from the old parties were well funded during campaigns. They were able to take out large advertisements in local newspapers, and they made frequent use of ready-made advertisements provided by the central organization. See, for example, *AG*, 9 October 1925, 3; 16 October 1925, 1; 23 October 1925, 2; *CPH*, 7 October 1925, 4; *PC*, 9 October 1925, 6; *FFP*, 8 October 1925, 4; *FS*, 8 October 1925, 4.

71 *AG*, 9 September 1921, 1; 16 September 1921, 1; *PC*, 30 September 1921, 4. Meighen trusted Stewart's political judgement, and in August 1921 he asked Stewart to assess the party's electoral chances in Lanark. Stewart replied that the riding of Leeds should be "disposed of" in a by-election before Lanark was contested in a general election. "If we do not win Leeds, the Liberals will, and in either event it will take the pep out of the farmers' ... movement in this locality." Once this was accomplished, winning Lanark would be easy for the Tories. NA, Meighen Papers, vol. 16, Stewart to Meighen, 23 August 1921; Meighen to Stewart, 24 August 1921.

72 *OP*, 27 October 1921, 5, 8. This was an important visit, in the view of W.A. Boys, South Simcoe's Conservative MP. Boys was disappointed that Meighen chose to speak in Orillia and not Barrie, even though Boys's "request was the first in." Boys was also displeased to learn that, despite his best efforts, he was not appointed to Meighen's cabinet before the election. He wrote to Meighen that "it was unfortunate no farmer was included in the cabinet ... it does seem to me that in view of the success of the UFO Movement and our desire to check it in rural Ridings, it would have been good policy to include a farmer, and in addition to that the importance of the agricultural industry, apart entirely from the [UFO], should warrant recognition." Boys was a farmer. NA, Meighen Papers, vol. 16, Boys to Meighen, 14 September 1921; Meighen to Boys, 16 September 1921; Boys to Meighen,

4 October 1921. See also the account of R.B. Bennett's visit to Stayner in Simcoe during the 1926 federal campaign. *CE*, 12 August 1926, 1, 4, 8.

73 *OT*, 14 June 1923, 6.

74 *PAT*, 15 October 1925, 1; *FS*, 15 October 1925, 1, 8; *FFP*, 15 October 1925, 1, 4. The bounty amounted to $75,000 that year. Greenizen thought that the amount was appropriate, given that $500,000 worth of oil was produced annually and between three and four million dollars were invested in the industry. Ibid. See also ibid., 22 October 1925, 2. H.J. Pettypiece reminded his readers that Fansher alone opposed the bounty. Ibid., 15 October 1925, 2; 29 October 1925, 2. Tory candidate J.E. Armstrong was also in favour of the bounty. *PAT*, 22 October 1925, 1; *FS*, 15 October 1925, 1.

75 Ibid., 27 January 1921, 1.

76 *PC*, 2 December 1921, 11. See also *AG*, 2 December 1921, 1, also 16 October 1925, 1; 23 October 1925, 1.

77 *OP*, 3 November 1921, 2. The alleged remarks actually had nothing to do with anything that Meighen had said. The comment was based on what a Toronto journalist overheard in a conversation between Manning Doherty, H.C. Nixon, and R.H. Grant, three Drury cabinet ministers. Grant wrote to Meighen to clarify the matter and denied ever saying that Meighen had called the Progressives bolshevists. Instead, he claimed that a member of Meighen's cabinet, Sir George Foster, had said that "the Farmers were Reds, or Bolsheviks." NA, Meighen Papers, Vol. 42, Grant to Meighen, 13 September 1921; Meighen to Grant, 14 September 1921.

78 *SFRN*, 1 September 1921, 1, 7; *AG*, 2 December 1921, 1; *CB*, 1 December 1921, 4.

79 *AG*, 8 June 1923, 1.

80 *OT*, 1 December 1921, 4.

81 Ibid., 24 November 1921, 6 (emphasis added). See also ibid., 1 December 1921, 1, 4; *OP*, 20 October 1921, 1; 3 November 1921, 11.

82 *PE*, 17 November 1921, 8; W.L. Morton, *The Progressive Party*, 154.

83 *PC*, 28 October 1921, 9; *AG*, 28 October 1921, 6.

84 *CB*, 1 December 1921, 5. Some UFO members were warned that the old parties would resort to such tactics. W.I. Johnson told Lanark UFO supporters in 1920 that in the next federal election the Union government would wave the flag and boast of its war record. He stated that "it was the producers of this country who had helped to win the war, not the Government at Ottawa at all." Johnson, it will be recalled, fought in the First World War. *SFRN*, 24 August 1920, 5. See also *AG*, 17 September 1920, 1.

85 *AG*, 24 November 1922, 6, 10.

86 Ibid., 1 December 1922, 5. The riding was won by Conservative R.F. Preston, who had served as MPP from 1894 to 1898 and from 1905 to 1919. He was minister without portfolio in the Whitney and Hearst administrations. Ontario, Chief Election Officer, *Candidates and Results*, 409; *SFRN*, 14 November 1922, 1. Preston also served as reeve of Carleton Place and then as that town's first mayor. Chambers, *The Canadian Parliamentary Guide 1923*, 183–4.
87 *OT*, 14 June 1923, 6.
88 *BE*, 8 October 1925, 1; *SUN*, 8 October 1925, 2; *CE*, 8 October 1925, 4. The Liberals could not bring themselves to support the Progressives; instead, they adopted an anti-Tory stance.
89 Goodison was a farmer and the president of the Goodison Thresher Co., which employed some 250 people. Thus, according to one supporter, he "had the interest of both the industrialist and agriculturalist at heart." *FFP*, 1 October 1925, 1, 4; *FS*, 29 October 1925, 1. It did Goodison little harm to have the support of *Free Press* editor H.J. Pettypiece, long known for his support of the UFO.
90 Palmer, *Capitalism*, 17.
91 *OT*, 9 February 1922, 4. For an earlier example of this sort of exercise, see *OP*, 27 February 1919, 7.
92 Ibid., 16 March 1922, 4. During the 1921 election "Politicus" had also decided to enlighten *Orillia Times* readers about agrarian protest. In two articles he misrepresented the Grange and POI and then presented a defence of the two-party system over group government: "For generations we have been governed by the two party system; it seems to suit *the disposition, temperament and genius of the Anglo Saxon race*." Ibid., 24 November 1921, 9 (emphasis added). See also ibid., 10 November 1921, 4.
93 Few justifications were put forward, possibly because it was taken for granted that what existed was the best of all possible worlds. See, for example, *CE*, 1 October 1925, 5.
94 *OT*, 2 March 1922, 4, 9 March 1922, 4. See also ibid., 9 August 1923, 4, for insights into liberalism, the system that stood for "individual responsibility, the most sacred thing on earth," a system that respected a man's sense of self, "and his rights to freedom of initiative and of development."
95 Muir supported PR, seeing it as a more democratic means of electing people to office. *AG*, 22 February 1923, 4. Ironically, on the same page that his editorial supporting PR was another one in which Muir praised Mackenzie King and his government for coming nearer "to the principles of true democracy than any government we have had." Muir made this claim after Progressive MP William Irvine had made a motion in the Commons that if a government measure was defeated,

it should not mean that the government should resign. King had replied that cabinet was the very basis of responsible government, and that no ruling party should be allowed to remain in power if defeated in a vote. To Muir, King's statement reached "the very foundations and basis of the British system of Constitutional Government." Thus, King, and not Irvine, was the keeper of the spirit of democracy.

96 Some people from the mainstream parties advocated changing the electoral system, but they were a decided minority. H.P. Hill, Tory MPP for Ottawa East and chair of the Legislative Committee that supported PR, spoke in glowing terms of the system to an Orillia audience. *OP*, 11 May 1922, 4; *OT*, 11 May 1922, 6. Even Mackenzie King supported PR, but predictably, he made no serious effort to implement it while in power. This still did not deter the partisan press from portraying him as a champion of democracy simply because he advanced a Redistribution Bill in 1924. *PC*, 25 July 1924, 2.

97 See, for example, *AG*, 14 March 1924, 3, for the article by F.A. Carman, "one of the most experienced of the Parliamentary press gallery correspondents." Carman perceptively noted that King held the Progressives hostage in the House of Commons with the threat of not passing the Redistribution Bill. He realized that the Progressives would not ally themselves with the Tories and defeat the government because the next election would then have to be fought using the existing ridings.

98 Fansher began diluting his platform in the 1925 campaign. See, for example, *FFP*, 22 October 1925, 1; 29 October 1925, 1. On his 1926 campaign, see *FS*, 26 August 1926, 2; 16 September 1926, 2; *FFP*, 16 September 1926, 2. West Lambton Progressives chose to support Liberal candidate William Goodison in 1926. Ibid., 2 September 1926, 2; 16 September 1926, 2; 25 September 1926, 4.

99 On the 1926 campaign see *FS*, 25 November 1926, 2, 6. Robert J. White contested Lambton West in 1926 but was defeated by Tory incumbent Wilfrid S. Haney. Ibid., 4 November 1926, 1; 11 November 1926, 1.

100 On his campaigns, see Drury, *Farmer Premier*, 166–71. Drury appears in election records as a Progressive, but as C.M. Johnston points out, by 1926 he had "shed the Progressive label and for all intents and purposes returned to the two-party fold as a 'free trade liberal.'" Johnston, *E.C. Drury*, 216.

101 Buchanan was a longtime UFO member and a buyer for the local UFCC. He promised to support a Liberal/Progressive government but reserved the right to vote independently when directed by his conscience (not his constituents) to do so. By that time, however, Buchanan was not interested in fighting the Big Interests; instead, he saw the central issue as the need to expand markets: "By scientific research and by still further vigorous action on the part of our Department of

Trade and Commerce these markets can be extended not only to the benefit of manufacturers but to the farmers as well." PC, 10 September 1926, 10. See also ibid., 27 August 1926, 4; AG, 20 August 1926, 1; 27 August 1926, 1; 10 September 1926, 5; 17 September 1926, 1; SFRN, 9 September 1926, 4; 14 September 1926, 1.

102 Prohibition Union candidate D.H. Gemmell was endorsed by the Lanark UFO Association in the 1926 provincial election, but, almost predictably, he withdrew shortly before election day. SFRN, 11 November 1926, 7; 18 November 1926, 1; PC, 12 November 1926, 3; 3 December 1926, 8; AG, 12 November 1926, 1. In a 1929 federal by-election the Lanark UFO voted against fielding a candidate. To do so, it was argued, would split the non-Conservative vote and ensure the Tories of a victory. *Almonte Gazette* editor James Muir was outraged at the decision, arguing that if Progressive "principles and policies were anything they were worth fighting for." He also argued that the UFO was alienating "many of [its] members who were formerly Conservatives by so frequently espousing the cause of the Liberal candidate in an effort to defeat the Conservatives." For Muir, the decision was an admission by local UFO members that the political side of the movement was dead. AG, 11 October 1929, 1–3.

103 Ontario, Chief Electoral Officer, *Candidates and Results*, J5-J6.

104 MacLeod, "The United Farmer Movement," 154–99.

CHAPTER FIVE

1 SUN, 18 September 1918, 6.

2 Ibid., 21 August 1918, 6. On the UFWO's formation and structure, see Kechnie, "United Farm Women," 267–9; Staples, 115–32. Women could join an existing UFO club or establish a UFWO club of their own.

3 Kechnie, "United Farm Women"; Rankin, "Politicization." Louise I. Carbert, in her *Agrarian Feminism*, relies on Kechnie's work and devotes only three pages to the UFWO.

4 Rankin, "Politicization," 310.

5 See, for example, Fink, *Agrarian Women*, 2, 23.

6 Griesbach was a school teacher from Collingwood. Raised on a farm, she edited and wrote most of the SUN's "Sisters' Page" from late 1917 to early 1922 when she was effectively fired by the SUN's editorial board, reportedly for her outspoken opinions.

7 SFRN, 11 November 1926, 7.

8 Griesbach herself saw the value of these letters: "The great majority of home-keeping women in the country have no way of really getting acquainted with their own mentality, except through self-expression in

letters. We have not the same opportunities for verbal expression as the men have." *SUN*, 17 April 1918, 6, also 4 February 1922, 6.

9 The UFWO (and the UFO) regularly addressed the subject of rural depopulation. This theme is discussed elsewhere in this study and in many secondary works on the UFO and consequently will be dealt with only peripherally here. For accounts of how local UFWO members felt about the subject, see *Sun*, 15 May 1920, 6; 2 February 1921, 6; 19 March 1921, 6; 25 June 1921, 6; 18 November 1926, 12; *FS*, 11 March 1926, 3.

10 *OP*, 3 May 1917, 4. Similar sentiments appeared elsewhere. See, for example, *Farmer's Advocate*, 53, no. 1,319 (3 January 1918): 4.

11 *SUN*, 15 May 1918, 6. See also the comments of "Kid" (ibid., 27 February 1918, 6), the letter from "Peggy" (ibid., 10 April 1923, 3), both from Simcoe, and ibid., 24 January 1920, 6. Fink argues that "Women in more marginal households were likely to be pressed into service as field workers and assigned a greater share of livestock chores than were women living on farms that could afford to hire workers." Fink, *Agrarian Women*, 52–3.

12 *SUN*, 1 May 1920, 6. See, as well, the letter from Alliston's "Ambrosine" (Ibid., 19 June 1923, 3), and the comments of Margery Mills in *SFRN*, 17 February 1920, 3. "Margery Mills" was a pseudonym used by Meta Schooly Laws of Cayuga, Ontario, who wrote a regular column in the *SUN*. Kechnie, "United Farm Women," 278, n. 14.

13 *SUN*, 5 June 1918, 6. See also Alice Webster's comments on farm women's work. Ibid., 26 March 1919, 6. Griesbach estimated that it took the average farm woman twelve hours a day to complete her household and farm tasks. Considering that leisure was relegated to a half-day on Sunday, the average work week totalled 78 hours. Ibid., 8 October 1921, 6. On the government's program to encourage city women to work on farms, see *FS*, 1 August 1918, 1.

14 In 1920 Griesbach observed that the bulletins of vocational information published by the Ontario Department of Labor made no mention whatsoever of homemaking as a profession. Griesbach estimated that this meant that the efforts of some 200,000 women were being ignored by the ministry. Noting that provincial Agriculture minister Manning Doherty had announced his intention to bring men to Ontario from Europe to help out on farms, she wrote, "I hope he will not overlook the fact that a large part of the burden of agriculture rests on the shoulders of farm women." Evidently, Doherty did overlook this fact. *SUN*, 14 August 1920, 6.

15 *SFRN*, 28 March 1922, 1, 4. A debate on the resolution that "farmers have more opportunity for pleasure than farm women" was held by

the Creemore UFO/UFWO in Simcoe in 1920. *SUN*, 27 March 1920, 6. On the inequality associated with purchases of labour-saving items on farms, see McGillvray, *Slopes of the Andes*, 59–60.

16 *OP*, 15 January 1920, 4; *SUN*, 8 May 1918, 6. Women UFO members in Smiths Falls built a restroom in that town even before the UFWO was established there. *SFRN*, 3 November 1920, 2; 23 December 1920, 1. A rest room had been constructed in Forest by the UFWO by mid-1922. *FS*, 29 June 1922, 3.

17 *SUN*, 12 March 1921, 6. Hydro was discussed at local meetings from time to time. See, for example, *SUN*, 21 August 1920, 6; 25 June 1921, 6. See also ibid., 14 June 1924, 3, for the views of the central UFWO. It is debatable, though, whether this was one of the main concerns of average farm women, as Kechnie, "United Farm Women," 270, implies.

18 *SUN*, 9 July 1921, 4; 31 December 1921, 6. See also her letter in ibid., 26 September 1922, 4. On rural as a historical category, see Walden, *Becoming Modern in Toronto*, 214.

19 *SUN*, 9 January 1918, 6.

20 Ibid., 14 January 1920, 6; 26 November 1921, 6. See also ibid., 2 April 1921, 6. Margery Mills occasionally wrote about press distortions. See, for example, ibid., 26 November 1919, 6. Griesbach herself was the victim of at least one fabrication by the press when the *Toronto Star* published a remark it attributed to her that farmers could sell milk for eight cents a quart and still make a reasonable profit. The *Star* eventually published a retraction, but by then the damage had been done. As Griesbach found out, many of the rural weeklies that carried the *Star* story failed to publish the retraction. *CE*, 8 January 1920, 7. On the issue of wartime press censorship, see *SUN*, 30 January 1918, 6.

21 Ibid., 24 September 1921, 6; 22 October 1921, 6. See also ibid., 10 September 1921, 6; 16 November 1921, 6.

22 Ibid., 14 June 1923, 3.

23 As Barbara Roberts notes, arguing for peace during peacetime is fairly uncontroversial, but "to promote peace in wartime, in contrast, is radical, certainly unpatriotic, and perhaps subversive or treasonous." Roberts, "Women's Peace Activism in Canada," 276. Activists could, however, also experience difficulties during peacetime for making 'objectionable' remarks. In 1925 outrage greeted Agnes Macphail when she stated that "even if Belgium had not been invaded, Britain would have found some excuse to enter the World War." *FS*, 27 August 1925, 1. Her remarks, made near Peterborough, were carried in papers throughout Ontario.

24 *SUN*, 14 February 1920, 6. The issue of war – its causes, its historical development, and the way to eliminate it – was the topic of a lesson for the United Farmers Young People of Ontario offered by the

Educational Department of the UFO. Referring to the First World War as "a crime against civilization" and recommending that students read Sir Philip Gibbs's *Now It Can Be Told (1920)*, the goal of the course was to "assist in achieving the ideal of universal peace." University of Guelph Library, Leonard Harman/UCO Collection, XA1, MS A126017, Lesson 1, "International Problems in their Relationship to World Peace." Agnes Macphail often wrote on the topic. See *SUN*, 8 September 1923, 5.

25 Ibid., 20 March 1920, 6; 11 December 1918, 6. See also ibid., 1 May, 1920, 6; 23 July 1921, 4.

26 Ibid., 2 April 1921, 6. Webster continued to write on the topic until at least 1926. See her letter in ibid., 16 September 1926, 4, in which she alluded to newspapers' vilifying farmers who went on record as abhorring warfare during the war. See also the comments of "Kid" of Simcoe County. Ibid., 27 February 1918, 6.

27 Ibid., 11 December 1918, 6; 21 August 1918, 6. On Church support of the war effort, see Allen, *Social Passion*, 35–45.

28 *SUN*, 20 March 1918, 6. Anti-militarism continued to be a concern of local UFWO members. See, for example, *CSN*, 27 February 1926, 2; 17 April 1926, 4; *CB*, 23 February 1928, 1; 13 June 1929, 2; 3 April 1930, 1.

29 *SUN*, 10 November 1920, 6.

30 Ibid., 22 May 1918, 6.

31 Ibid., 12 December 1917, 6.

32 Ibid., 14 January 1920, 6; 31 January 1920, 6; 14 February 1920, 6. See also Kechnie, "United Farm Women," 276.

33 *SUN*, 31 January 1920, 6. On Webster's thoughts on community laundries, see ibid., 22 October 1921, 6.

34 Ibid., 7 April 1920, 6.

35 *CSN*, 26 June 1920, 1, 3.

36 *SUN*, 3 July 1920, 6. "Bachelor" of Lambton County once wrote to Griesbach and chided her for ignoring letters from "mere" men. Griesbach replied that, since she was committed to the equality of men and women, she deprecated the use of the word "mere" to describe men: "Be assured … that we are glad to have our 'brother man' show a disposition to exchange views with us." Ibid., 7 May 1919, 6.

37 *AG*, 23 June 1922, 1. Interestingly, the correspondent covering the event described this section of Macphail's speech as "strange."

38 Ibid., 20 May 1927, 6.

39 *SUN*, 10 April 1923, 3.

40 Ibid., 2 April 1925, 13.

41 Ibid., 27 February 1918, 6.

42 Ibid., 26 March 1921, 6. Collins had been drawn to patent medicines because she was attempting to avoid surgery for an unmentioned

condition. It is interesting to note that the bunion specialist who duped her claimed to have been "sent out by the Government." The tactic of claiming government accreditation was often used by swindlers. See Badgley, "'Then I saw I had been swindled,'" 342–3.

43 SUN, 12 June 1920, 6.

44 Ibid.

45 Ibid., 12 June 1920, 6; 14 January 1922, 6.

46 Ibid., 12 February 1921, 6.

47 Ibid., 2 July 1921, 4. See also ibid., 12 December 1917, 6; 5 June 1918, 6; 18 September 1918, 6; 4 December 1918, 6; 31 January 1920, 6; 20 March 1920, 6; 18 June 1921, 6; 2 July 1921, 4.

48 Ibid., 24 April 1920, 6; 15 May 1918, 6. "Louise" was likely Louise Collins, a frequent SUN contributor, who also referred to herself occasionally as "Sister Lou." Her allusion to her time as not being as valuable as that of others in the family may have been a tongue-in-cheek remark, or another example of the tendency of women to deprecate their work.

49 Cohen, *Women's Work*, 92–117. See also Derry, "Gender Conflicts in Dairying."

50 PC, 20 April 1917, 3. See also the comments made during a debate held by the Smiths Falls UFO. SFRN, 28 March 1922, 1, 4. See also FS, 1 May 1921, 3; 28 February 1924, 8.

51 SUN, 20 March 1918, 6.

52 AG, 16 January 1920, 2.

53 SUN, 14 August 1920, 6.

54 AG, 24 March 1922, 1; CPH, 29 March 1922, 5.

55 Ibid., 7 March 1924, 1. For other evidence, see "Minutes of the Fourth Annual Convention of the United Farm Women of Ontario 1922," and "Report on Marketing from the Fifth Annual Convention in 1923." Universit of Guelph Library, UCO, XA1 MSA126005. See also SUN, 18 June 1921, 6; 18 December 1924, 13.

56 AG, 14 March 1924, 1. Griesbach commented from time to time on women's control of milk, poultry, and small fruit production on Ontario farms. On one occasion, she added that, when she mentioned women in these activities, she was referring to "average Ontario farms … not a joint-stock company … with an immense amount of capital involved." SUN, 14 August 1920, 6.

57 Ibid., 9 January 1918, 6, also 2 October 1920, 6; 26 June 1918, 6. On the perishable nature of women's produce, see Cohen, 116.

58 There were, however, ways in which to exercise some control in the price women received for their produce. Louise Collins half-jokingly announced her family's intention in the coming summer to "have a baseball game with the eggs we can't use. That might be one way of

breaking up the chain of stores lined up to say just what we farmers ought to get for eggs this summer." SUN, 25 March 1922, 6.

59 Ibid., 4 December 1918, 6.

60 AG, 2 December 1921, 7.

61 SUN, 26 November 1919, 6. Collins had written to Griesbach even before women got the vote. Commenting on the results in her riding for the 1917 federal election, she noted that Liberal candidate E.C. Drury had lost to a "selfish, greedy trickster who hands out political patronage royally with one hand, and with the other gathers in every sort of personal gain and advantage." The campaign had not increased Collins's "respect for male suffrage." Ibid., 26 December 1917, 6.

62 Ibid., 15 October 1921, 6.

63 Ibid., 4 August 1920, 6. Earlier that year, Collins had written to Diana that she had noticed several men drinking alcohol on election day. She supposed it was necessary for them to "drown their conscience and vote for their party right or wrong." Ibid., 24 January 1920, 6.

64 Ibid., 19 July 1919, 6.

65 AO, RG 4–32 1918, file 1711. See also Griesbach's comments on government controls on free speech. SUN, 1 May 1920, 6.

66 OSN, 30 August 1919, 1. Brodie was one of the founding members of the UFWO and was its first president. Staples, *The Challenge*, 119.

67 OT, 16 October 1919, 5. Generally, the same can be said of Agnes Macphail and Griesbach during the 1921 federal election. See CE, 6 October 1921, 1; OT, 17 November 1921, 3; OP, 17 November 1921, 9. Griesbach's main speech was entitled "Shall Big Interests Rule?"

68 CB, 13 November 1919, 11. For examples of temperance matters being discussed at local meetings, see AG, 21 March 1924, 8, 28 March 1924, 1; FS, 10 April 1924, 3. Griesbach wrote very little on the subject. For a rare exception, see SUN, 16 April 1921, 6.

69 SUN, 16 July 1921, 6.

70 OP, 24 November 1921, 1; OT, 24 November 1921, 6. McNabb also spoke on behalf of East Simcoe UFO candidate J.B. Johnston in the 1923 provincial election. Ibid., 14 June 1923, 5.

71 CB, 6 October 1921, 1.

72 SUN, 22 April 1922, 6, 26 November 1921, 6; Rankin, "Politicization," 317. On Macphail, see CE, 6 October 1921, 1; 17 August 1922, 4; FS, 6 July 1922, 4,; AG, 25 April 1924, 3; 20 May 1927, 6; 22 June 1928, 1; BNA, 24 July 1924, 2; OP, 25 October 1923, 1; CB, 26 November 1925, 7.

73 PC, 11 November 1921, 4.

74 AG, 17 September 1920, 1. See also Johnson's comments at Smiths Falls earlier that year. SFRN, 24 August 1920, 5. See also SUN, 10 July 1918, 7, for similar sentiments expressed at a Forest UFO picnic. For a

discussion of the widespread idea that women were morally elevated, see Osterud, *Bonds of Community*, 72–80.

75 *AG*, 15 July 1921, 7. In the 1925 federal election, UFWO members in North Simcoe acted as codirectors for the townships in that riding. *CB*, 1 October 1925, 1.

76 *SFRN*, 22 November 1921, 8; *AG*, 4 November 1921, 1.

77 Actually, token gestures were made to women in the 1919 provincial election. See, for example, ibid., 10 October 1919, 4. But even the UFO tended to ignore, at least initially, the importance of securing women supporters. At the meeting to select a UFO candidate for South Lanark in 1919, no women were present. Moreover, the only person in attendance at the meeting who is reported to have lamented this was A.A. Powers, president of the UFO Publishing Co. *PE*, 2 October 1919, 3.

78 *OT*, 24 November 1921, 1, 6.

79 *AG*, 28 October 1921, 1; *PE*, 17 November 1921, 8.

80 *SFRN*, 1 December 1921, 7. Stewart was rumoured to be a Progressive candidate in a 1922 Lanark federal by-election, although nothing came of this. Ibid., 31 October 1922, 1.

81 Ibid.; *PE*, 20 October 1921, 4.

82 *SUN*, 3 September 1921, 6; 26 November 1921, 6.

83 *AG*, 19 May 1922, 1.

84 Ibid., 24 March 1922, 1; *CPH*, 29 March 1922, 5.

85 *SUN*, 27 March 1920, 6; *CB*, 15 November 1928, 8.

86 *SUN*, 11 March 1926, 9.

87 Ibid., 15 December 1923, 3; University of Guelph Archives, UCO, XA1 MS A126005, "Minutes of the Fifth Annual Convention of the United Farm Women of Ontario 1923."

88 On the Women's Institute, see Ambrose, *For Home and Country*, "'What Are the Good'"; Howes, "Adelaide Hoodless"; and Crowley, "Madonnas before Magdalenes."

89 Kechnie, "United Farm Women," 269.

90 *SUN*, 3 July 1918, 6.

91 Ibid., 22 April 1922; 19 June 1918, 6. See also AG, 16 November 1928, 1.

92 By the early 1920s the *SUN* was providing accounts of the annual Women Institute meeting. *SUN*, 20 November 1924, 2. In addition, in 1921 the UFWO passed a resolution that it be represented at all future Women's Institute conventions, and that the institute be represented at all future UFWO conventions. As was reported at that meeting, "In many sections there has been a marked improvement in the relationship between the WI and the UFWO. We hope before the coming year is out to evolve a plan whereby these two [groups] shall realize that they can accomplish the ends for which both are working best by cooperative effort. University of Guelph Archives, UCO, XA1

MS A126005, "Minutes of the Third Annual Convention of the United Farm Women of Ontario 1921."

93 *SUN*, 24 September 1921, 7. In Uhthoff, Ardtrea, and Price's Corners, farmers' clubs and the institute met jointly before the UFO established itself in the area. See *OP*, 21 December 1916, 4; 28 December 1916, 6; 4 January 1917, 8; 1 March 1917, 4; 5 April 1917, 6; 31 May 1917, 6. After the farmers' clubs became affiliated with the UFO, farm women continued to meet under the auspices of the institute. See ibid., 16 February 1922, 4; *OT*, 19 June 1919, 3. The same also seems to apply to the UFO in Edenvale, where a UFWO does not appear to have been formed. *BNA*, 2 March 1922, 4.

94 *OT*, 3 August 1922, 6. In other cases, it appears that the UFWO club actually led to a decline of a an institute club. In Lanark, the Carleton Place Women's Institute disbanded in 1928, while at the same time the local UFWO club enjoyed continued popularity. *AG*, 13 July 1928, 1. On the popularity of the Carleton Place UFWO, see ibid., 30 March 1928, 7. On the continued popularity of the UFWO in Lanark, see ibid., 23 October 1931, 1; 13 November 1931, 3.

95 See, for example, *OT*, 15 March 1923, 7.

96 *SUN*, 29 January 1925, 8; *FS*, 11 March 1926, 3.

97 *SFRN*, 11 November 1926, 7; *SUN*, 19 March 1925, 8. Donations to causes was a regular feature of institute work. See, for example, *AG*, 20 August 1915, 1; 7 July 1922, 6.

98 *SUN*, 7 May 1925, 8.

99 Rankin, "Politicization," 318.

100 *SUN*, 19 February 1925, 8.

101 Ibid., 9 July 1925, 8. See also the accounts of the Union Club 604A in *CB*, 19 July 1928, 2, and of the Carleton Place UFWO in *CPH*, 5 March 1930, 1.

102 See, for example, the letter written by "Centre Simcoe." (*SUN*, 14 June 1923, 3) and the two letters by Louise Collins (ibid., 29 May 1923, 3; 19 June 1923, 3).

103 Ibid., 21 July 1923, 5.

104 *AG*, 18 December 1925, 3; *SUN*, 15 September 1927, 3.

105 Ibid., 3 November 1922, 1. That same year the Kinburn UFO, near Lanark, elected two women to its seven-person executive. *AG*, 6 January 1922, 3. Gardner was still secretary in 1924 and was married in that year to David Hollie Lowry. Lowry was from Ramsay and owned a house on the ninth line, close to where Gardner's family lived. Gardner resigned her position as schoolteacher in Ramsay shortly before her marriage. Ibid., 11 July 1924, 1.

106 *SFRN*, 11 November 1926, 7; *AG*, 26 October 1928, 1. As well, Lanark did not fit the pattern of most counties in that women were not

always responsible (as they were elsewhere) for overseeing the local United Farm Young People's Organization (UFYPO). From 1927 to 1929 M.B. Cochran assumed this role, and in 1931 Fred McTavish took on the task. *SFRN*, 11 November 1926, 7; *PC*, 4 November 1927, 1; *AG*, 26 October 1928, 1.

107 NA, W.C. Good Papers, Webster to Good, 4 February 1922. See also Webster to Good, 20 March 1922.

108 See Kechnie, "United Farm Women," 275–6.

CHAPTER SIX

1 *CSN*, 13 September 1919, 5.

2 McGillivray and Ish, *Co-operatives in Principle and Practice*, 5.

3 Fairbairn et al., "Co-operative Institutions," 33.

4 Goodwyn, *Democratic Promise*, xviii, and *Populist Moment*, 66.

5 See, for example, Laycock, *Populism and Democratic Thought*, 28–9; Wylie, "Direct Democrat," 193–4.

6 See Wood, *A History*, 73–90, 118–20.

7 Despite the extensive literature on the role of the state, no thorough account of its role in encouraging, supporting, and shaping the development of Canadian cooperatives has been written. In the few studies that do mention the state, some view it in neutral or positive terms. See Shey, "Co-operative Marketing of Agricultural Products in Ontario"; Mooney, *Co-operatives Today and Tomorrow*; Perkins, *Co-operatives in Ontario*. Others are ambiguous as to the effects of state involvement. See McGillivray and Ish, *Cooperatives in Principle*; Macpherson, "Creating Stability Amid Degrees of Marginality"; Laycock, *Co-operative-Government Relations in Canada*. I found only a handful of studies that highlighted the problems associated with state intervention into cooperative activity. See Davidovik, *Towards a Co-operative World*, 48–53, and Harrison, *Modern State*, 190.

8 See Booth, "Agricultural Cooperation in Canada," 357. On pre-WWI federal policy see Fowke, *Canadian Agricultural Policy*, 188–250.

9 *BNA*, 17 May 1923, 1; O'Brien, "The Cooperative Marketing of Fleece Wool in Canada"; *The Golden Fleece*. By 1945 the company controlled some sixty-five percent of Canada's annual wool clip. Canada, *Report of the Royal Commission on Co-operatives*, 176.

10 AO, Ferguson Papers, Box 60, File 03–06–0–366 1925, "Status of Cooperative Organizations in Canada."

11 NA, RG 17, vol. 2984, File 32–2, Live Stock Commissioner to J.H. Grisdale, 27 July 1920; vol. 2958, File 30–5–2 pt. 1, Live Stock Commissioner to J.H. Grisdale, 27 August 1921. On the importance of Canada's bacon industry, see *AG*, 13 October 1922, 2; Ontario, *Report of the Agricultural Enquiry Committee 1924*, 34.

12 *AG*, 24 June 1920, 1. Federal authorities also distributed a great deal of literature. In 1920 the Department of Agriculture's list of publications contained some 350 titles.

13 Ibid., 14 January 1916, 4; *OP*, 28 March 1918, 1; 17 December 1915, 1; 11 June 1924, 1; *CPH*, 18 January 1916, 1. See also the account of the provincially operated "Better Farming Special" railway (ibid., 23 November 1915, 6). On government demonstrations, see *SFRR*, 27 December 1917, 5.

14 *OSA*, 16 March 1916, 1; *FS*, 16 October 1919, 4; 7 August 1922, 7; *AG*, 19 October 1923, 1. The province also sponsored the establishment of Boys' and Girls' Live Stock Clubs with the aim of promoting the use of purebred stock and providing an incentive for staying on the farm. It also devised mechanisms by which local clubs could obtain the credit needed to purchase the animals. *OT*, 17 March 1921, 10.

15 *SFRR*, 28 February 1918, 1; Wylie, "Direct Democrat," 217–19.

16 *SUN*, 9 May 1917, 3. See also the Conservative party's 1919 provincial election pamphlet, *An Agricultural Policy that is Efficient: Leadership that Means Service and Stability* (AO, Pamph. 1919, no. 102), 3.

17 Later, the Drury government also attempted to secure farm help from urban areas, ostensibly to assist with the harvest but also to try to reduce urban unemployment. *FS*, 30 December 1920, 3.

18 Space limitations do not permit a full discussion of the campaign, but accounts of typical Patriotism and Production meetings can be found in *AG*, 5 March 1915, 1; *SUN*, 17 February 1915, 3. For campaign propaganda, see Canada, Department of Agriculture, *Production and Thrift*; *FS*, 3 May 1917, 7; *AG*, 11 May 1917, 7.

19 AO, RG 3, Box 60, File 03–06–0–366 1925, "Status of Cooperative Organizations in Canada"; *FS*, 25 November 1920, 4. This program obtained less-than-stellar results. See Ontario, *Report of the Agricultural Enquiry Committee 1924*, 18.

20 *SUN*, 25 July 1917, 8.

21 *SFRN*, 18 July 1922, 5. The same article was also located in *FS*, 20 July 1922, 5; *OT*, 3 August 1922, 6; *AG*, 30 June 1922, 1; *CPH*, 2 August 1922, 6; *BNA*, 20 July 1922, 5. For examples of such columns, see *AG*, 4 April 1919, 5; 7 January 1921, 5; 9 June 1922, 4; 19 September 1926, 6; *CB*, 22 May 1919, 8; *BNA*, 4 September 1924, 2; *FS*, 8 May 1919, 8; 25 November 1920, 4; 6 January 1921, 6; 17 September 1924, 5. It is a rare occasion to search through smalltown papers from that period and *not* see an article provided by the Department.

22 State officials tended to blame farmers for many problems. See *SFRR*, 1 February 1917, 3; *CPH*, 25 June 1918, 3.

23 *PC*, 17 October 1919, 11 (emphasis added).

24 Ibid. See also the advertisement for Lanark's Tory candidate (J.A. Stewart) in the 1921 federal election. *AG*, 28 October 1921, 6. For other

examples, see ibid., 10 September 1926, 3; OT, 1 December 1921, 4; CB, 9 September 1926, 2. See also Bristow, "Agrarian Interest," 117.

25 On Sapiro's tour of the Prairies in 1923–24, see MacPherson, *Building and Protecting*, 77.

26 PC, 10 March 1922, 3; 31 March 1922, 3; AG, 23 June 1922, 2. Implicit throughout this study is the notion that the state underwent no fundamental change as a result of the election of the UFO. As J.E. Hodgetts argues (*From Arm's Length*, 188), although one might have expected some change in the provincial bureaucracy as a result of Drury becoming premier, "the inexperience that led to debilitating scandals, and the constant tension between elected members of the Farmers' party and strident extra-parliamentary groups led by J.J. Morrison, all combined to frustrate any dramatic changes in past practices."

27 Ontario, *Report of the Agricultural Enquiry Committee 1924*, 5, 17, 34.

28 Ibid., 70. The Ontario government's activities in the area of cooperation culminated in the creation of a provincial agricultural marketing board in 1931. Significantly, one of the original members of the board was H.B. Clemes, manager of the UFCC. FS, 30 April 1931, 1. Perkins (*Cooperatives in Ontario*, 38), while sympathetic to state intervention in cooperatives, nevertheless notes that marketing boards, unlike co-ops, are not voluntary and are "monopolistic bodies which seek to raise prices to producers through bargaining power, discriminatory pricing ... [and] have extensive powers to control the marketing of their designated product."

29 Ontario, *Report of the Agricultural Enquiry Committee, 1924*, 91. UFCC president George A. Bothwell also referred to this subsidy at the company's annual meeting in 1924. BNA, 18 December 1924, 4.

30 Robertson, *Some Occurrences and Conditions Overseas*, 15–19.

31 Changing consumer demands in early-twentieth-century Ontario meant a call for greater standardization in and attractiveness of product. As Ontario Agriculture officials noted, "In the old days the wormy apple, the misshapen potato and the old hen may not have lost their attractiveness, but times have changed ... the demands for foods that appeal to the eye and to the sense of taste have increased very greatly." AG, 9 June 1922, 4. Evenness of product was also important: "The certainty in the mind of the English bacon curer that he can always get a similarity of product in his purchases is apparently the main reason why Danish bacon ... always commands a wholesale price from two to four dollars more a hundredweight over Canadian." OT, 9 August 1923, 4. See also FS, 20 March 1919, 2; AG, 14 March 1924, 1; Ontario, *Report of the Agricultural Enquiry Committee 1924*, 36.

32 One contemporary observer, George Keen of the Cooperative Union of Canada (CUC), was suspicious about cooperative marketing

associations, believing that they could become vehicles for class privilege. He also maintained that such groups depended on the efficiency of their leaders for success, rather than on their individual members. See Skey, "Cooperative Marketing," 59–60; MacPherson, *Building and Protecting*, 37. Despite these misgivings, the CUC admitted such bodies as members because "genuine producers such as marketing cooperatives are organized to eliminate the element of profit on price, and instead to substitute reward for actual service and that it is desirable that both types should ... work together for their mutual advantage as against economic interests which operate to their common injury." Keen, *The History of the Co-operative Movement in Canada*, 5–6.

33 MacPherson, *Each for All*, 29–33. The general treatment of consumer cooperatives may be attributed to complaints lodged by retail merchants to both federal and provincial authorities soon after the formation of the UFCC. Skey, "Cooperative Marketing," 48.

34 Fowke, "Canadian Farmers in an Industrial Society," 2–4.

35 *SUN*, 14 April 1915, 3; 8 February 1917, 1; 3 February 1915, 3; 15 March 1915, 3; *OSA*, 9 March 1916, 1, 11 February 1915, 1; *FS*, 1 June 1916, 2; 17 August 1916, 3; *PT*, 5 April 1916, 1. In 1915 Lambton was the first county in Ontario to replace the Farmers' Institutes with a Board of Agriculture, the object of which was to further the interests of agriculture through information dissemination, to encourage local talent, and to act as an umbrella organization for all agricultural associations in the county. The idea of creating boards was devised by the provincial government in response to the declining popularity of the Farmers' Institutes. They did not, however, meet with much success. For the case of Lambton, see *PT*, 22 September 1915, 7; 17 November 1915, 1; *OSA*, 18 November 1915, 5. Boards were also set up in Simcoe and managed to survive until 1923, although in greatly reduced form. AO, RG 16, Series G-5–1, MS 597, Reel 56, "Simcoe North 1915–1916," 32; "Simcoe North, 1921–22," 14. In Lanark, the scheme failed miserably. Ibid., Reel 30, "Lanark 1917–1918," 12; "Lanark 1918–1919," 16.

36 *OSA*, 1 April 1915, 1; *PT*, 31 March 1915, 1. This was a constant problem for the UFCC. See Skey, "Cooperative Marketing," 53; *The United Farmers' Co-operative Digest*, 1, no. 5 (December 1922): 3. Binder twine was one of the first products offered to farmers by the UFO on a cooperative basis, and it was not without its share of controversy. The UFCC originally attempted to secure the twine from the Canadian Cordage Co. but was refused. An arrangement was then struck between the UFCC and an Irish firm, causing many to think that farmers were disloyal for not buying Canadian-made goods. *SFRN*, 5 August 1920, 1; Skey, "Cooperative Marketing," 36–8. Later, UFO members felt compelled to defend the actions of their brethren on the

Prairies when they were denounced as unpatriotic for buying American-made implements. As UFO faithful explained it, the Grain Growers Grain Company (GGGC) was forced to buy from American firms because Canadian manufacturers refused to do business with it. *OT*, 17 November 1921, 3–4; 24 November 1921, 4.

37 *OSA*, 1 April 1915, 1.

38 *SUN*, 3 February 1915, 3; *PT*, 12 May 1915, 3. In all three counties (and throughout the province) there was a constant state presence in the form of agricultural representatives. Relations between the representatives and the UFO in Lambton and Simcoe were cordial, but in Lanark relations became increasingly strained to the point where the representative spoke at very few UFO meetings and complained of being "looked upon as an outsider." Lanark UFO clubs eventually refused to submit figures associated with their cooperative enterprises (see the representative's reports for the 1919–26 period, AO, RG16). There is no evidence to indicate why this tension between the local representative and Lanark UFO members existed, but it may be another example of the greater militancy seen in Lanark UFO supporters. Soon after the Tories regained provincial power in 1923, North Simcoe representative Alan Hutchinson was asked to resign his post, allegedly for campaigning for the Progressives in the 1921 federal election. Hutchinson denied the charge and speculated that it was his help in organizing a farmers' cooperative potato company that led to his difficulties. Agriculture minister J.B. Martin claimed Hutchinson was dismissed for "inefficiency." A number of local papers and farmers' clubs rushed to his defence but to little avail. *OT*, 6 December 1923, 2; Simcoe County Archives, Rugby Minutes, December 1923. On the formation of the company, see Hutchinson's report for 1921–22, AO, RG16, 34.

39 *OSA*, 13 May 1915, 5; *SUN*, 14 April 1915, 3; *PT*, 31 March 1915, 1; 12 January 1916, 4.

40 Ibid., 19 January 1916, 1; *OSA*, 20 January 1916, 5. No evidence could be found to suggest that such an option was available to clubs belonging to the UFCC.

41 *PT*, 21 December 1916, 1.

42 Ibid., 5 April 1917, 1. On Groh's removal from the UFCC executive, see Wylie, "Direct Democrat," 201–2.

43 *SUN*, 14 February 1917, 4; *PT*, 4 July 1917, 5. On cooperative potato shipments, see ibid., 17 May 1917, 1.

44 *SUN*, 16 January 1918, 3. Noble as this policy may have been, it had a negative impact on local cooperatives because there was no incentive to remain a member. Later, the Oro Station UFO in Simcoe County reversed a similar policy on the grounds that it was too difficult to

collect money from non-members. Simcoe County Archives, Oro Minutes, 2 August 1923. See also MacPherson, *Each for All*, 315.

45 *PT*, 24 May 1917, 1.

46 *SUN*, 2 April 1919, 10; *PAT*, 14 October 1920, 2.

47 AO, RG 16, Series G-5–1, MS 597, Reel 56, "Simcoe North 1916–1917," 25; "Simcoe North, 1917–1918," 19–21.

48 *SUN*, 17 July 1918, 8. In addition to this activity, a cooperative society was formed in Orillia in 1919 by George Keen. *OT*, 6 November 1919, 3; 20 November 1919, 6.

49 *CB*, 26 June 1919, 1.

50 *SUN*, 18 June 1919, 10.

51 AO, RG 16, Series G-5–1, MS 597, Reel 56, "Simcoe North 1918–1919," 20. According to the agricultural representative, most of the business conducted by these clubs involved shipping produce, although clubs did purchase salt, sugar, molasses, etc. on a cooperative basis. *CSN*, 29 March 1919, 1. Around that time the Uptergrove UFO made its first cooperative shipment of livestock to Toronto. *OP*, 13 March 1919, 6. As well, a number of farmers' clubs not formally affiliated with the UFO were shipping and purchasing cooperatively. See, for example, the account of the Ardtrea farmers' club in ibid., 28 November 1918, 8.

52 *OT*, 13 July 1916, 1.

53 Ibid., 19 July 1917, 5; *OP*, 12 July 1917, 6. For examples of UFO leadership visiting other clubs, see *FS*, 22 June 1916, 2; *PC*, 13 July 1917, 8; *SUN*, 12 June 1918, 3.

54 *AG*, 11 December 1914, 1, 17 December 1915, 1; 26 January 1917, 4; AO, RG 16, Series G-5–1, MS 597, Reel 30, "Lanark 1913–1914," 8–9; On the club's formation, see ibid., "Lanark 1908–1910," 1.

55 *AG*, 11 February 1916, 8.

56 Ibid., 26 March 1915, 4.

57 Ibid., 14 December 1917, 1. It might be noted that, as was the case in Simcoe, the representative also felt it necessary to castigate the farmers for not keeping proper accounts. Forsyth in effect blamed farmers for any losses they may have incurred. It seems likely that he provided at least some of the arguments against cooperative buying, especially in light of the fact that he seldom made reference to consumer cooperatives in his annual reports, and those comments that were made strongly suggest that he was, at best, lukewarm to the concept.

58 Ibid., 28 February 1919, 4.

59 Ibid., 21 March 1919, 4.

60 *PC*, 20 April 1917, 1.

61 Ibid.

62 Ibid. (emphasis added).

63 AO, RG 16, Series G-5–1, MS 97, Reel 30, "Lanark 1918–1919", 16–8; *SUN*, 11 June 1919, 10. The Balderson UFO club attained success even earlier than some of the clubs mentioned here. By early 1917 the club had over one hundred members and had conducted over $20,000 in business during the previous year. By late 1917 the membership had grown to over 120. *PC*, 20 April 1917, 4; 7 September 1917, 8.
64 *SFRR*, 26 August 1919, 4. See also *SUN*, 2 April 1919, 10; 11 June 1919, 10.
65 Ibid., 11 June 1919, 10; *SFRR*, 26 August 1919, 4; *SFRN*, 14 October 1919, 4. Hitchcock was hired by the UFCC, but he was likely nominated by the local club, as was common practice. See Skey, "Cooperative Marketing," 31.
66 *SFRN*, 17 February 1921, 4. From what was revealed in Powers's report, the $3,000 initially stated as being in the bank may have been an exaggeration, although some of those funds may have been expended by the time Powers audited the store. The difficulty that the UFCC had managing branch stores from Toronto did not mean that efforts to do so were not made. In 1921 the UFCC passed a motion to instruct store managers that they were "not to make any purchases without the approval of the Manager of the Branch Store Department, with the exception of fruit … fish, meat, produce, and oil or gasoline." University of Guelph Library, UCO, XA1 MS A126009, File 2, Minutes, 22 February 1921. By this time all pricing was done at head office. Wylie, "Direct Democrat," 221. As Skey observes ("Cooperative Marketing," 31), local members were consulted in management matters from time to time, "but the ultimate control rested with the Central."To be fair, some of the UFCC leadership favoured centralized operations because, if any local store experienced a loss, then it would not be borne by local members but instead would be distributed among all stores. In other words, successful stores would carry the less-successful ones.
67 *BNA*, 14 December 1922, 10.
68 Many other explanations were advanced regarding the failure of the stores. It was argued that the prices charged by the stores offered no incentive for customers to patronize them. When first established, the stores undercut local merchants. This led, however, to merchants undercutting UFO store prices, or to ill will on the part of merchants and manufacturers. So strong were these pressures that it was not long before the UFCC began charging the general retail price for groceries. McCutcheon, "Economic Organization," 62; Skey, "Cooperative Marketing," 61. Farmers were then to be paid "patronage dividends" at year's end. Some farmers, reluctant to wait a year before receiving a share of the profits (if there were any profits), shopped at other stores. Wylie, "Direct Democrat," 223. UFCC president R.J. Scott admitted in

1938 that "perhaps the most serious mistake was that the stores were owned and managed from the central instead of having each owned and controlled as an independent enterprise by the members who patronized it." Scott, "Co-operative Purchasing in Ontario," 379.

69 *AG*, 23 November 1923, 7; Wylie, "Direct Democrat," 194.
70 See, for example, the account of the founding meeting of the Lanark Village UFO club. *LE*, 26 November 1919, 1.
71 *PC*, 24 December 1920, 3.
72 *CPH*, 17 November 1920, 1; AO, RG 16, Series G-5-1, MS 597, "Lanark 1921–22," 16.
73 *SFRN*, 24 August 1920, 5. Later that year Willoughby castigated some farmers for taking the bait offered by drovers and forsaking the UFO stock shippers. "There are some farmers who would sell their birth-right for ten cents; they don't see any farther than their noses." Ibid., 3 November 1920, 2.
74 Ibid., 11 January 1921, 1. It seems that the club took almost every activity very seriously. A special meeting was held in March 1921 to discuss – heatedly as it turned out – which seed to buy cooperatively. Ibid., 15 March 1921, 5.
75 The mill had a capacity of two hundred barrels per day. Ibid., 13 May, 1920, 1. On the purchase of the mill, see ibid., 25 March 1920, 1; *CPH*, 31 March 1920, 1.
76 University of Guelph Library, Minutes, 29 March, 1920; 31 May 1920; 5 July 1920.
77 Ibid., 5 July 1920; 10 July 1920.
78 *SFRN*, 15 June 1920, 4. As seen earlier, Lanark was ill suited for wheat growing. Later, C.E. Merkley, a UFCC director for Eastern Ontario, also criticized local farmers for not sending more wheat to the mill. Ibid., 1 February 1923, 4.
79 NA, W.C. Good Papers, vol. 14, 10427–10431.
80 University of Guelph Library, Minutes, 22 February 1921.
81 *SFRN*, 3 March 1921, 3; University of Guelph Library, Minutes, 22 November 1922. The Smiths Falls mill lost $4,877 in 1922. *SFRN*, 7 December 1922, 3.
82 University of Guelph Library, Minutes, 1 February 1923 (emphasis added). The UFCC was able to extricate itself from the five-year "lease" after only three years because no formal agreement was ever signed. The UFCC took possession of the mill and acted according to the terms of a draft lease, which was never officially consummated. Ibid.
83 Ibid., 1 March 1923. The mill was sold to the Smiths Falls Water Commission for a reported $35,000. *SFRN*, 22 February 1923, 1; *PC*, 11 May 1923, 1.

84 OP, 26 August 1920, 5; 14 October 1920, 2. Even before the Smiths Falls purchase the Port Perry UFO club established a company capitalized at $60,000 in order to operate the Carnegie flour, saw, and planing mills. SUN, 3 October 1917, 4. As well, the farmers' clubs of Howard Township (Kent County) bought the evaporator and flour mill at Ridgetown in 1919. FS, 12 June 1919, 1. And in 1920 the Brant Farmers' Cooperative Society purchased the local Dominion Flour Mills building. Wylie, "Direct Democrat," 245. Later, the prospect of purchasing controlling interest in the Chelsey Woollen Mills for some $40,000 was discussed and apparently acted upon by UFO members in that area. BNA, 1 September 1921, 7.

85 FS, 8 January 1920, 1; 13 January 1921, 1.

86 Dairying and poultry were the primary objects of state support, but efforts were also made in areas such as livestock and wool. SFRN, 1 February 1923, 1; AG, 10 August 1923, 4; 31 October 1924, 1; PC, 21 April 1922, 3; annual reports of the agricultural representative for the period 1922–35.

87 SFRN, 18 March 1920, 2.

88 CPH, 3 November 1919, 1. See also AG, 7 November 1919, 1; AO, RG 16, Series G-5-1, MS 597, Reel 30, "Lanark 1918–19," 1.

89 Reprinted in PC, 2 July 1920, 3.

90 AO, RG 16, Series G-5-1, MS 597, Reel 30, "Lanark," 1919–20, 21, 1920–21, 12, and 1921–22, 16.

91 See AG, 14 March 1924, 1; CPH, 27 May 1925, 4. A similar scheme was attempted in Simcoe. AO, RG 16, Series G-5-1, MS 597, Reel 56, "Simcoe North,", 1923–24, 12; BNA, 14 February 1923, 1; OP, 21 February 1924, 5, 8.

92 See AO, RG 16, Series G-5-1, MS 597, Reel 31, "Lanark," for the years 1923–24 (11), 1924–25 (12), 1925–26 (12), and 1927–28 (15), for descriptions of the Middleville, Balderson, and Harper egg circles, all of which prospered well into the 1930s.

93 PC, 29 September 1922, 7; AG, 19 January 1923, 8. The failure to attain its goals may have been due, in part, to the creation of the Ontario Milk and Cream Producers in 1923, another provincewide organization, "fostered by the provincial Ministry of Agriculture." Ibid., 30 March 1923, 7.

94 Ibid., 27 February 1931, 7. The federal government granted thirty percent of the costs, while the provincial government donated at least thirty-five percent of the funds. The story of the dairy industry in Lanark is more complicated than it seems here. Many firms often tried to sway farmers to patronize their creameries and cheese factories over others. See, for example, PC, 25 June 1926, 4; AG, 3 September 1926, 1; 11 March 1927, 8; 25 September 1927, 1.

95 *FS*, 18 May 1922, 1. See also ibid., 20 January 1921, 1; 4 May 1922, 5. On the formation of the club, see *SUN*, 12 January 1921, 7. The contract system was also used by celery growers in the Thedford district. Ibid., 8 June 1922, 1. On UFCC contracts, see Skey, "Cooperative Marketing," 85–6. Relations with private firms continued to be a problem. Agricultural representatives frequently gave the names of UFO club secretaries to private firms so that they could approach them with better deals. University of Guelph Library, Minutes, 22 February 1921.

96 *FS*, 22 June 1922, 1.

97 Ibid., 18 January 1924, 6; 22 May 1924, 6. The club continued to be successful long after the UFO ceased to exist. Ibid., 5 June 1924, 1; 22 January 1925, 4; 29 January 1931, 8; Amos, *The Wanstead Co-op Story 1924–1974*.

98 *FS*, 24 June 1920; 1 July 1920, 2; 11 November 1920, 3. A store was established in Thedford as well. Ibid., 9 December 1920, 6; 6 January 1921, 2.

99 Ibid., 1 May 1924, 3; 18 February 1926, 3. The UFCC Egg and Poultry Department reorganized in 1924, largely because many of its members complained that the only difference from past practice was that the net earnings accrued to the UFCC instead of to private buyers. After 1924, patronage dividends were paid to members, as opposed to offering better prices, which had been the practice prior to the reorganization. Perkins, *Cooperatives in Ontario*, 50.

100 *FS*, 10 February 1927, 1. By late July 1927 8,000 farmers in the province had done so, representing some hundred thousand acres of wheat. By 1928 the number had increased to 9,600. Ibid., 28 July 1927; 5 April 1928, 8. In 1928 a livestock pool was formed along similar lines but, owing to difficulties at the UFCC, it was discontinued after 1929. Skey, "Cooperative Marketing," 88; *FS*, 5 January 1928, 1.

101 Ibid., 1 March 1928, 1.

102 AO, RG 16, Series 16.09, Box A46, File "Lambton Growers' Cold Storage Ltd." See also *FS*, 19 March 1931, 2; 26 March 1931, 2; 31 May 1931, 1; 14 July 1932, 1.

103 *SUN*, 21 August 1920, 4; University of Guelph Library, Minutes, 10 July 1920.

104 *BNA*, 28 April 1921, 4; 24 February 1921, 4.

105 Ibid., 14 February 1923, 1.

106 *CB*, 22 January 1925, 8; 16 April 1925, 8. Ibid., 21 January 1926, 8; 27 January 1927, 4. See also Simcoe County Federation of Agriculture, *Report*, 14–5.

107 *CB*, 15 March 1928, 2; Perkins, 43–4.

108 See *OP*, 12 August 1920, 5; 8 June 1922, 2; 17 July 1924, 5.

109 CB, 19 May 1921, 2. See also AO, RG 16, Series G-5–1, MS 597, Reel 56, "Simcoe North," for the years 1921–22 to 1926–27.
110 OT, 20 July 1920, 3.
111 Yet cooperation remained popular. As late as 1932 there were over 135 co-op associations in Ontario, with roughly 40,000 members. Skey, "Cooperative Marketing," 223.
112 University of Guelph Library, UCO, XA1 MS A126017, lessons 4, 6, 7, 8. The quotation is from lesson 8 (emphasis added).
113 An indication of the movement towards full support for cooperative marketing can be found in "Report on Cooperative Marketing – 1925," written for the UFCC by Mrs J.S. Amos. Ibid., XA1 MS A126008, File 5. See also Skey, "Cooperative Marketing," 63, and the account of a UFO co-op marketing campaign in Lanark (AG, 7 March 1924, 1). On earlier efforts, see SUN, 10 September 1921, 8; 4 July 1922, 3. On the travelling UFO co-op marketing course, see FS, 30 January 1930, 2; 20 February 1930, 2.
114 Ibid., 17 March 1922, 2; 22 June 1922, 1; AG, 14 March 1924, 1; PC, 17 March 1922, 6; SUN, 13 August 1921, 7.
115 Simcoe County Archives, acc. 977–10 E18 B1 R3B S1 sh1 (United Farmers of Ontario and UF Co-op), A.J. Saunders to UFO club secretaries, 15 January 1927, 23 March 1927; Skey, 53.
116 Laycock, *Populism and Democratic Thought*, 34.
117 This, of course, assumes that the group wished to form a new cooperative. Often, it was easier to join one of the many that existed already. In fact, one of the problems farmers faced was that there were too many cooperatives from which to choose.

CHAPTER SEVEN

1 E.E. Cummings, *A Miscellany Revised*, 327.
2 SUN, 14 January 1921, 6.
3 OT, 17 November 1921, 3.
4 Wylie, "Direct Democrat," 419. By 1926 many UFO clubs in Lanark experienced decreases in membership. In that year the Perth and Almonte UFO clubs reported losses in membership and the Smiths Falls UFO "met so seldom that the members had nearly lost track of one another." SFRN, 11 November 1926, 7.
5 The Edenvale UFO reached its peak with thirty-two members in 1920. By 1924 twelve members remained. In Oro Station, membership peaked in 1919, when the club had twenty-three members. Membership fell to sixteen by 1926. The Rugby UFO had twenty-two members in 1920, but by 1926 the number had fallen to fourteen.

Simcoe County Archives, Edenvale Minutes, Rugby Minutes, and Oro Minutes.

6 At the 1995 meeting of the Canadian Association for Studies in Cooperation, Charles A. Diemer, an Essex County farmer, presented a moving defence of the cooperative ethic. He became involved in cooperatives in the late 1930s and at times his words bore striking similarities to those of UFO members. Diemer sounded even more like some of the more radical UFO members when he concluded, "There is absolutely no possible way to attain lasting world peace, the end to violence and all those other social bewilderments by leaving it to elected representatives in government. It is essential for all of us to accept the challenge as both a moral responsibility as well as a privilege to cooperate in a profound way and help to write the next chapter of history." Diemer, "Changes in Cooperatives," 7.

7 Although not within the scope of this study, it bears noting that, during the period in which it held power, the UFO introduced some important legislation that attempted to assist the weak and powerless. However much one wishes to characterize the legislative accomplishments of the UFO as mildly social democratic or as accommodating to the advancement of modern capitalism, it remains the case that many of the laws the UFO government passed were ahead of most North American jurisdictions at the time. See Johnston, *E.C. Drury,* 150–65.

8 Susan Brown, "Anarchism, Feminism, Liberalism and Individualism," 23.

Bibliography

PRIMARY SOURCES

Archival Records

ARCHIVES OF ONTARIO

RG 3, Premier's Papers

Series 5, E.C. Drury Papers

Series 6, G. Howard Ferguson Papers

RG 4, Attorney General

RG 16, Department of Agriculture and Food, "Annual Reports of Agricultural Representatives 1907–1969, MS 597, Reel 30, "Lanark 1921–22."

MU 7778, Varney, Ontario: Grange/United Farmers of Ontario Minutes, 1911–1925

LAMBTON COUNTY ARCHIVES

Box 16(2), File 16A-F, Minutes, East Lambton UFO

NATIONAL ARCHIVES OF CANADA

MG 26 I, Arthur Meighen Papers

MG 27 III C 1, W.C. Good Papers

RG 17, Department of Agriculture

SIMCOE COUNTY ARCHIVES

Minute Book of the Oro Station UFO

Minute Book of the Rugby UFO
Minute Book of the Edenvale UFO

UNIVERSITY OF GUELPH LIBRARY
Leonard Harman/UFO Collection
XA1 MS A126009, Minutes of the Meeting of the Board of Directors and the United Farmers Co-operative Co. Ltd.
XA1, MS A126017, Minutes of the United Farm Woman of Ontario

Newspapers

LANARK COUNTY
Almonte Gazette
Carleton Place Herald
Lanark Era
Perth Courier
Perth Expositor
Rideau Record (Smiths Falls)
Smiths Falls Record-News

LAMBTON COUNTY
Forest Free Press
Forest Standard
Oil Springs Advance
Petrolia Advertiser-Topic
Petrolia Topic
Sarnia Canadian Observer

SIMCOE COUNTY
Barrie Examiner
Collingwood Bulletin
Collingwood Enterprise
Collingwood Saturday Bulletin
Collingwood Saturday News
Northern Advance (Barrie)
Orillia Packet
Orillia Packet and Times
Orillia Saturday News
Orillia Times

OTHER NEWSPAPERS
Farmer's Advocate
Weekly Sun (Farmers' Sun)

Government Publications

Canada. *Report of the Royal Commission on Co-operatives.* Ottawa: King's Printer, 1945.

– Department of Agriculture. *Production and Thrift: Agricultural War Book 1916.* Ottawa: King's Printer. 1916.

Census of Canada for 1911, 1921, 1931.

Ontario. *Report of the Agricultural Enquiry Committee 1924.* Toronto: King's Printer, 1925.

– *Report of the Committee on Proportional Representation.* Toronto: Legislative Assembly of Ontario, 1921.

– Department of Agriculture. *Annual Report of the Statistics Branch 1920.* Toronto: King's Printer, 1921.

– Department of Agriculture and Food. *Origins, Classification and Use of Ontario Soils.* Toronto: Department of Agriculture and Food, 1974.

– Chief Election Officer. *Candidates and Results: With Statistics from the Records 1867–1982.* Toronto: Chief Election Officer, n.d.

– *Centennial Edition of a History of the Electoral Districts, Legislatures and Ministries of the Province of Ontario 1867–1967.* Toronto: Chief Election Officer, 1967.

Contemporary Works

H. Belden & Co. *Illustrated Historical Atlas of the County of Lambton.* First published 1880. Reprinted with additions 1973. Sarnia: E. Phelps.

– *Historical Atlas of Lanark and Renfrew Counties.* First published 1880–81. Reprinted 1972. Owen Sound: Richardson, Bond & Wright.

– *Historical Atlas of Simcoe County.* First published 1881. Reprinted Port Elgin, Cumming Atlas Reprints, 1975.

Ernest J. Chambers, ed. *The Canadian Parliamentary Guide 1923.* Ottawa: Mortimer, 1923.

Ontario Liberal-Conservative Party. *An Agricultural Policy that Is Efficient: Leadership that Means Service and Stability.* Toronto: N.p., 1919.

The United Farmers' Co-operative Digest. Toronto, 1923.

SECONDARY SOURCES

Allen, Richard. *The Social Passion: Religion and Social Reform in Canada 1914–28.* Toronto: University of Toronto Press, 1973.

Ambrose, Linda M. *For Home and Country: The Centennial History of the Women's Institutes in Ontario.* Toronto: Boston Mills Press, 1996.

– "'What Are the Good of Those Meetings Anyway?': Early Popularity of the Ontario Women's Institutes." *Ontario History* 87, no. 1 (March 1995): 1–19.

Amos, Helen E. *The Wanstead Co-op Story 1924–1974*. N.p., 1974.

Anderson, Jim, Isaac Applebaum, Jane Craig, Ilka Karl, Robert Sweeny, Jamie Swift, Harmiena van Oosten, and Robert Winslow. *A Political History of Agrarian Organizations in Ontario 1914–1940: With Special Reference to Grey and Bruce Counties*. N.p.: Chase Press, 1973.

Angus, James T. *A Deo Victoria: The Story of the Georgian Bay Lumber Company, 1871–1942*. Thunder Bay: Severn Publications, 1990.

Armstrong, Christopher, and Nelles, H.V. *Monopoly's Moment: The Organization and Regulation of Canadian Utilities, 1830–1930*. Toronto: University of Toronto Press, 1988.

Badgley, Kerry A. "The Social and Political Thought of the Farmers' Institute, 1884–1917: Manifestations of Agrarian Discontent." MA thesis, Carleton University, Ottawa, 1988.

– "Ringing out the Narrowing Lust of Gold, Ringing in the Common Love of Good: The United Farmers of Ontario in Lambton, Simcoe, and Lanark Counties, 1914–1926." PhD thesis, Carleton University, Ottawa, 1996.

– "'Then I saw I had been swindled': Frauds and Swindles Perpetrated on Farmers in Late-Nineteenth-Century Ontario." In *Canadian Papers in Rural History Vol. IX*, edited by Donald H. Akenson. Gananoque: Langdale Press, 1994.

Bakunin, Michael. *Bakunin on Anarchy: Selected Works by the Activist-Founder of World Anarchism*. Edited, translated, and introduced by Sam Dolgoff. New York: Alfred A. Knopf, 1972.

Beaven, Brian P.N. "Partisanship, Patronage, and the Press in Ontario, 1880–1914: Myths and Realities." *Canadian Historical Review* 64, no. 3 (September 1983): 317–51.

Bennett, Carol. *In Search of Lanark*. Renfrew: Juniper Books, 1982.

Biesenthal, Linda. *A Rural Legacy: The History of the Junior Farmers' Association of Ontario*. N.p.: Junior Farmers' Association of Ontario, 1981.

Bittermann, Rusty "The Hierarchy of the Soil: Land and Labour in a 19th Century Cape Breton Community." *Acadiensis* 17, no. 1 (Autumn 1988): 33–55.

Bitterman, Rusty, Robert A. MacKinnon, and Graeme Wynn. "Of Inequality and Interdependence in the Nova Scotia Countryside, 1850–70." *Canadian Historical Review* 74, no. 1 (March 1993): 1–43.

Bland, Warren R. "The Location of Manufacturing in Southern Ontario in 1881." *Ontario Geography*, no. 8 (1974): 8–39.

Bliss, Michael. *A Canadian Millionaire: The Life and Business Times of Sir Joseph Flavelle, Bart. 1858–1939*. Toronto: Macmillan, 1978.

– *Northern Enterprise: Five Centuries of Canadian Business*. Toronto: McClelland and Stewart, 1987.

Bothwell, Robert. *A Short History of Ontario*. Edmonton: Hurtig, 1986.

Bristow, Dudley Alexander. "Agrarian Interest in the Politics of Ontario: A Study with Special Reference to the Period 1919–1949." MA thesis, University of Toronto, 1950.

Brown, Howard Morton. *Lanark Legacy: Nineteenth Century Glimpses of an Ontario County.* Perth: Corporation of the County of Lanark, 1984.

Brown, L. Susan. "Anarchism, Liberalism and Individualism," *Our Generation* 24, no. 1 (Spring 1993): 22–61.

– *The Politics of Individualism: Liberalism, Liberal Feminism and Anarchism.* Montreal: Black Rose Books, 1993.

Brown, Wayne Crawford. "The Broadening Out Controversy: E.C. Drury, J.J. Morrison and the United Farmers of Ontario." MA thesis, University of Guelph, 1979.

Byers, Daniel T. "The Conscription Election of 1917 and its Aftermath in Orillia, Ontario." *Ontario History* 83, no. 4 (December 1991): 275–96.

Canadian Society of Technical Agriculturalists. *Agricultural Cooperation in Canada.* Ottawa: Canadian Society of Technical Agriculturalists, 1938.

Chomsky, Noam. *The Chomsky Reader.* Edited by James Peck. New York: Pantheon Books, 1987.

– *Toward a New Cold War: Essays on the Current Crisis and How We Got There.* New York: Pantheon Books, 1982.

– *For Reasons of State.* New York: Pantheon Books, 1973.

Chomsky, Noam, and Edward S. Herman. *Manufacturing Consent: The Political Economy of the Mass Media.* New York: Pantheon Books, 1988.

Clark, John. *The Anarchist Moment: Reflections on Culture, Nature and Power.* Montreal: Black Rose Books, 1984.

Clarke, Harold D., Jane Jenson, Lawrence LeDuc, Jon H. Pammett. *Political Choice in Canada.* Toronto: McGraw-Hill Ryerson, 1979.

Clifford, N. Keith. *The Resistance to Church Union in Canada 1904–1939.* Vancouver: University of British Columbia Press, 1985.

Cohen, Marjorie Griffin. *Women's Work, Markets, and Economic Development in Nineteenth-Century Ontario.* Toronto: University of Toronto Press, 1988.

Conway, John. "'To Seek a Goodly Heritage': The Prairie Populist Resistance to the National Policy." PhD thesis, Simon Fraser University, Vancouver, 1979.

Cook, Ramsay. "Tillers and Toilers: The Rise and Fall of Populism in Canada in the 1890s." *Historical Papers 1984 Communications historiques.* Ottawa: Canadian Historical Association, 1985.

Cooper, Barbara J. "Farm Women: Some Contemporary Themes." *Labour/Le Travail* 24 (Fall 1989): 167–80.

Craig, John. *Simcoe County: The Recent Past.* N.p.: Corporation of the County of Simcoe, 1977.

Craig, John G. *The Nature of Cooperation.* Montreal: Black Rose Books, 1993.

Crewison, Daryll, and Ralph Matthews. "Class Interests in the Emergence of Fruit-Growing Co-operation in Lincoln County, Ontario, 1880–1914." In *Canadian Papers in Rural History Vol. V*, edited by Donald H. Akenson. Gananoque: Langdale Press, 1986.

Crowley, Terry. "Agnes Macphail and Canadian Working Women." *Labour/ Le Travail*, no. 28 (Fall 1991): 129–48.

– *Agnes Macphail and the Politics of Equality.* Toronto: James Lorimer and Company, 1990.

– "Madonnas before Magdalenes: Adelaide Hoodless and the Making of the Canadian Gibson Girl." *Canadian Historical Review* 67, no. 4 (1986): 520–47.

– "The New Canada Movement: Agrarian Youth Protest in the 1930s." *Ontario History* 80, no. 4 (December 1988): 311–25.

Davidovik, G. *Towards a Cooperative World: Economically, Socially, Politically.* Antigonish: Coady International Institute, 1967.

Davis, Angela E. "'Country Homemakers': The Daily Lives of Prairie Women as Seen through the Woman's Page of the Grain Growers Guide 1908–1928." In *Canadian Papers in Rural History Vol. VIII*, edited by Donald H. Akenson. Gananoque: Langdale Press, 1992.

Derry, Margaret. "Gender Conflicts in Dairying: Ontario's Butter Industry, 1880–1920." *Ontario History* 90, no. 1 (Spring 1998): 31–47.

Diemer, Charles A. "Changes in Cooperatives – Changes in Society and Culture." Unpublished paper presented at the annual meeting of the Canadian Association for Studies in Cooperation, Montreal, 1995.

Doucet, Michael, and John Weaver. *Housing the North American City.* Kingston: McGill-Queen's University Press, 1991.

Drummond, Ian. *Progress without Planning: The Economic History of Ontario from Confederation to the Second World War.* Toronto: University of Toronto Press, 1987.

Drury, E.C. *Farmer Premier: The Memoirs of E.C. Drury.* Toronto: McClelland and Stewart, 1966.

Elford, Jean Turnbull. *Canada West's Last Frontier: A History of Lambton.* Sarnia: Lambton County Historical Society, 1982.

– *A History of Lambton County.* Sarnia: N.p., 1967.

English, John. *The Decline of Politics: The Conservatives and the Party System 1901–20.* Toronto: University of Toronto Press, 1977.

Evans, A. Margaret. *Sir Oliver Mowat.* Toronto: University of Toronto Press, 1992.

Fairbairn, Brett, June Bold, Murray Fulton, Lou Hammond Ketilson and Daniel Ish. *Cooperatives and Community Development: Economics in Social Perspective.* Saskatoon: Centre for the Study of Cooperatives, 1991.

Fink, Deborah. *Agrarian Women: Wives and Mothers in Rural Nebraska, 1880–1940.* Chapel Hill: University of North Carolina Press, 1992.

Finkel, Alvin. "Populism and Gender: The UFA and Social Credit Experience." *Journal of Canadian Studies* 27, no. 4 (Winter 1992–93): 76–97.

Fisher, David Hackett. *Historians' Fallacies: Toward a Logic of Historical Thought*. New York: Harper and Row, 1970.

Fowke, Vernon. *Canadian Agricultural Policy: The Historical Pattern*. Toronto: University of Toronto Press, 1946. Reprinted 1978.

– "Canadian Farmers in an Industrial Society." Unpublished paper, Barrie, 1964.

Fox-Genovese, Elizabeth. "From Separate Spheres to Dangerous Streets: Postmodernist Feminism and the Problem of Order." *Social Research* 60, no. 2 (Summer 1993): 107–23.

Fulton, Murray E. *Cooperative Organizations and Canadian Society: Popular Institutions and the Dilemmas of Change*. Toronto: University of Toronto Press, 1990.

Gagan, David. *Hopeful Travellers: Families, Land, Social Change in Mid-Victorian Peel County, Canada West*. Toronto: University of Toronto Press, 1981.

– "Enumerator's Instructions for the Census of Canada 1852 and 1861." *Histoire social/Social History* 7, no. 14 (November 1974): 353–66.

– "Writing the History of Ontario in the 1980s: Defining a Distinctive Society." *Acadiensis* 21, no. 1 (Autumn 1991): 166–79.

Ghorayshi, Parvin. "Canadian Agriculture: Capitalist or Petit Bourgeoise?" *Canadian Review of Sociology and Anthropology* 24, no. 3 (1987): 358–73.

Good, W.C. *Farmer Citizen: My Fifty Years in the Canadian Farmers' Movement*. Toronto: Ryerson Press, 1958.

Goodwyn, Lawrence. *Democratic Promise: The Populist Moment in America*. New York: Oxford University Press, 1976.

– *The Populist Moment: A Short History of the Agrarian Revolt in America*. New York: Oxford University Press, 1978.

Graham, W.H. *Greenbank: Country Matters in 19th Century Ontario*. Peterborough: Broadview Press, 1988.

Gray, Edward C., and Barry E. Prentice. "Exploring the Price of Farmland in Two Ontario Localities Since Letters Patenting." In *Canadian Papers in Rural History Volume IV*, edited by Donald H. Akenson. Gananoque: Langdale Press, 1984.

Griezic, Foster J.K. "An Introduction." In Louis Auubrey Wood. *A History of Farmers Movements in Canada*. Toronto: University of Toronto Press, 1975. First published 1924.

– "'Power to the People': The Beginning of Agrarian Revolt in Ontario: the Manitoulin By-Election, 24 October 1918." *Ontario History* 69, no. 1 (March 1977): 33–54.

Guerin, Daniel. *Anarchism*. Introduction by Noam Chomsky. Translated by Mary Klopper. New York: Monthly Review Press, 1970.

Hallowell, Gerald A. *Prohibition in Ontario, 1919–1923*. Ottawa: Ontario Historical Society, 1972.

Hanam, H.H. *Pulling Together for 25 Years: A Brief Story of Events and People in the United Farmers' Movement in Ontario During the Quarter Century, 1914–1939*. Toronto: United Farmers of Ontario, 1940.

Hann, Russell. *Farmers Confront Industrialism: Some Historical Perspectives on Ontario Agrarian Movements*. Toronto: New Hogtown Press, 1975.

Harrison, Frank. *The Modern State: An Anarchist Analysis*. Montreal: Black Rose Books, 1983.

Heick, W.H. *A Propensity to Protect: Butter, Margarine and the Rise of Urban Culture in Canada*. Waterloo: Wilfrid Laurier Press, 1991.

Hillman, W.B. "J.J. Morrison: A Farmer Politician in an Era of Social Change." MA thesis, University of Western Ontario, London, 1974.

Hodgetts, J.E. *From Arm's Length to Hands On: The Formative Years of Ontario's Public Service, 1867–1940*. Toronto: University of Toronto Press, 1995.

Hoffman, J.D. "Farmer-Labour Government in Ontario." MA thesis, University of Toronto, 1959.

Humphries, Charles W. *'Honest Enough to Be Bold': The Life and Times of Sir James Pliny Whitney*. Toronto: University of Toronto Press, 1985.

Hunter, Andrew F. *A History of Simcoe County*. Barrie: Historical Committee of Simcoe County, 1909.

Irvine, William. *The Farmers in Politics*. Toronto: McClelland and Stewart, 1976. First published 1920.

Hoffman, David. "Intra-Party Democracy: A Case Study." *Canadian Journal of Economics and Political Science* 27, no. 2 (May 1961): 223–35.

Horowitz, Gad. "Toward the Democratic Class Struggle." *Journal of Canadian Studies* 1, no. 3 (November 1966): 3–10.

Howes, Ruth. "Adelaide Hoodless." In *The Clear Spirit*, edited by Mary Q. Innis. Toronto: University of Toronto Press, 1966.

Jahn, Cheryle. "Class, Gender and Agrarian Socialism: The United Farm Women of Saskatchewan, 1926–1936." *Prairie Forum* 19, no. 2, (Fall 1994): 189–206.

Johnson, J.K., ed. *The Canadian Directory of Parliament 1867–1967*. Ottawa: Public Archives of Canada, 1968.

Johnson, J.K., and Bruce Wilson, eds. *Historical Essays on Upper Canada: New Perspectives*. Ottawa: Carleton University Press, 1989.

Johnston, Charles M. *E.C. Drury: Agrarian Idealist*. Toronto: University of Toronto Press, 1986.

– "'A Motley Crowd': Diversity in the Ontario Countryside in the Early Twentieth Century." In *Canadian Papers in Rural History Vol. VIII*, edited by Donald H. Akenson. Gananoque: Langdale Press, 1990.

Kalmakoff, Elizabeth. "Naturally Divided: Women in Saskatchewan Politics, 1916–1919." *Saskatchewan History* 46, no. 2 (Fall 1994): 3–18.

Katz, Michael. *The People of Hamilton, Canada West: Family and Class in a Mid-Nineteenth-Century City.* Cambridge, MA: Harvard University Press, 1975.

Kealey, Gregory S. "State Repression of Labour and the Left in Canada, 1914–20: The Impact of First World War." *Canadian Historical Review* 73, no. 3 (September 1992): 281–314.

– "The Writing of Social History in English Canada, 1970–1984." *Social History* 10, no. 3 (October 1985): 347–65.

Kealey, Gregory S., and Bryan D. Palmer. *Dreaming of What Might Be: The Knights of Labor in Ontario, 1880–1900.* Toronto: New Hogtown Press, 1987. First published 1982.

Kechnie, Margaret. "The United Farm Women of Ontario: Developing a Political Consciousness." *Ontario History* 77, no. 4 (December 1985): 267–80.

Keen, George. *The History of the Cooperative Movement in Canada.* Brantford: Cooperative Union of Canada, 1943.

Kelly, Kenneth. "The Development of Farm Produce Marketing Agencies and Competition Between Market Centres in Eastern Simcoe County, 1850–1875." In *Canadian Papers in Rural History Vol. I,* edited by Donald H. Akenson. Gananoque: Langdale Press, 1976.

Keshen, Jeff. "All the News That Was Fit to Print: Ernest J. Chambers and Information Control in Canada, 1914–19." *Canadian Historical Review* 73, no. 3 (September 1992): 315–43.

Kulikoff, Allan. "The Transition to Capitalism in Early America." *William and Mary Quarterly* 46 (1989): 121–49.

Laycock, David. *Cooperative-Government Relations in Canada: Lobbying, Public Policy Development and the Changing Co-operative System.* Saskatoon: Centre for the Study of Cooperatives, 1987.

– *Populism and Democratic Thought in the Canadian Prairies, 1910 to 1945.* Toronto: University of Toronto Press, 1990.

– *Prairie Populists and the Idea of Cooperation, 1910–1945.* Saskatoon: Centre for the Study of Cooperatives, 1985.

Lauriston, Victor. *Lambton's Hundred Years 1849–1949.* Sarnia: Haines Frontier, 1949.

Leigh, Jabez Montgomery. *Monty Leigh Remembers: Early Days in Oro Township and Orillia,* edited by Grace Leigh and Sally Gower. Orillia: Simcoe County Historical Society, 1983.

Leithner, Christian. "The National Progressive Party of Canada, 1921–1930: Agricultural Economic Conditions and Electoral Support." *Canadian Journal of Political Science* 26, no. 3 (September 1993): 435–53.

Levy, Andrea. "Progeny and Progress: Reflections on the Legacy of the New Left." *Our Generation* 24, no. 2 (Fall 1993-Winter 1994): 1–38.

Lockwood, Glenn J. *Beckwith: Irish and Scottish Identities in a Canadian Community.* Carleton Place: Township of Beckwith, 1991.

MacDonald, Donald C. "Ontario's Political Culture: Conservatism with a Progressive Component." *Ontario History* 86, no. 4 (December 1994): 297–317.

MacGillivray, Royce. *The Mind of Ontario.* Belleville: Mika Publishing, 1985.

– *The Slopes of the Andes: Four Essays on the Rural Myth in Ontario.* Belleville: Mika Publishing, 1990.

MacLeod, Jean. "The United Farmer Movement in Ontario, 1914–1943." MA thesis, Queen's University, Kingston, 1958.

MacPherson, C.B. *The Political Theory of Possessive Individualism: Hobbes to Locke.* Toronto: Oxford University Press, 1979. First published 1962.

MacPherson, Ian. *Building and Protecting the Cooperative Movement: A Brief History of the Cooperative Union of Canada.* Ottawa: Cooperative Union of Canada, 1984.

– "The Cooperative Union of Canada and Politics, 1909–31." *Canadian Historical Review* 54, no. 2 (June 1973): 152–74.

– "Creating Stability Amid Degrees of Marginality: Divisions in the Struggle for Orderly Marketing in British Columbia 1900–1940." In *Canadian Papers in Rural History Vol. VII*, edited by Donald H. Akenson. Gananoque: Langdale Press, 1990.

– *Each for All: A History of the Cooperative Movement in English Canada, 1900–1945.* Toronto: Macmillan of Canada, 1979.

– "An Authoritative Voice: The Reorientation of the Canadian Farmers' Movement, 1935 to 1945." In *Historical Papers 1979* Ottawa: Canadian Historical Association, 1980.

MacPherson, Ian, and John Herd Thompson. "An Orderly Reconstruction: Prairie Agriculture in World War Two." In *Canadian Papers in Rural History Vol. IV*, edited by Donald H. Akenson. Gananoque: Langdale Press, 1984.

– "The Business of Agriculture: Prairie Farmers and the Adoption of 'Business Methods,' 1880–1950." In *Canadian Papers in Business History Vol. I*, edited by Peter Baskerville. Victoria, BC: Public History Group, 1989.

McCalla, Douglas. *Planting the Province: The Economic History of Upper Canada.* Toronto: University of Toronto Press, 1993.

McCutcheon, Wilfred Whyte. "Economic Organization and the Development of the United Farmers Cooperative Company, Limited." MSA thesis, University of Toronto, 1948.

McGillivray, Anne, and Daniel Ish. *Cooperatives in Principle and Practice.* Saskatoon: Centre for the Study of Cooperatives, 1992.

McInnis, R. Marvin. *Perspectives on Ontario Agriculture 1815–1930.* Gananoque: Langdale Press, 1992

McKay, Ian. "Gazing through the Quintland Window: The Ethics of Dionnology?" *Journal of Canadian Studies* 29, no. 4 (Winter 1994–95): 144–52.

Marr, William. "Did Farm Size Matter? An 1871 Case Study." In *Canadian Papers in Rural History Vol. VI*, edited by Donald H. Akenson. Gananoque: Langdale Press, 1988.

Melnyk, George. *The Search for Community: From Utopia to a Cooperative Society.* Montreal: Black Rose Books, 1985.

Menzies, Heather. *By the Labour of Their Hands: The Story of Ontario Cheddar Cheese.* Kingston: Quarry Press, 1994.

Mika, Nick, and Helma Mika. *Places in Ontario: Their Names, Origins and History.* Belleville: Mika Publishing, 1981.

Millbrath, Lester W. *Political Participation: How and Why Do People Get Involved in Politics?* Chicago: Rand McNally, 1965.

Mooney, George S. *Cooperatives Today and Tomorrow: A Canadian Survey.* Montreal: Survey Committee, 1938.

Morris, Brian. *Bakunin: The Philosophy of Freedom.* Montreal: Black Rose Books, 1993.

Morton, Desmond. "*Sic Permanent*: The People and Politics of Ontario." In *The Government and Politics of Ontario 2d Edition*, edited by Donald C. MacDonald. Toronto: Van Nostrand Reinhold, 1980.

Morton, Desmond, and Glenn Wright. *Winning the Second Battle: Canadian Veterans and the Return to Civilian Life 1915–1930.* Toronto: University of Toronto Press, 1987.

Morton, W.L. *The Progressive Party in Canada.* Toronto: University of Toronto Press, 1950. Reprinted with corrections 1967.

Naylor, James. *The New Democracy: Challenging the Social Order in Industrial Ontario, 1914–1925.* Toronto: University of Toronto Press, 1991.

Noel, S.J.R. *Patrons, Clients, Brokers: Ontario Society and Politics 1791–1896.* Toronto: University of Toronto Press, 1990.

O'Brien, Allan. "The Cooperative Marketing of Fleece Wool in Canada," MSC thesis, Cornell University, Ithaca, 1950.

Oliver, Peter. *G. Howard Ferguson: Ontario Tory.* Toronto: University of Toronto Press, 1970.

– *Public and Private Persons: The Ontario Political Culture 1914–1934.* Toronto: Clarke, Irwin, 1975.

Osterud, Nancy Grey. *Bonds of Community: The Lives of Farm Women in Nineteenth-Century New York.* Ithaca: Cornell University Press, 1991.

Palmer, Bryan D. *Capitalism Comes to the Backcountry: The Goodyear Invasion of Napanee.* Toronto: Between the Lines, 1994.

– *Descent into Discourse: The Reification of Language and the Writing of Social History.* Philadelphia: Temple University Press, 1990.

– "The Eclipse of Materialism: Marxism and the Writing of Social History in the 1980s." In *Socialist Register 1990*, edited by Ralph Miliband, Leo Panitch and John Saville. London: Merlin Press, 1990.

– *Working Class Experience: Rethinking the History of Canadian Labour, 1800–1991.* Toronto: McClelland and Stewart, 1992.

Panitch, Leo, ed. *The Canadian State: Political Economy and Poliical Power.* Toronto: University of Toronto Press, 1977.

Parr, Joy. *The Gender of Breadwinners: Women, Men, and Change in Two Industrial Towns 1880–1950*. Toronto: University of Toronto Press, 1990.

Parsons, Stanley B., Karen Toombs Parsons, Walter Killilae, and Beverly Borgers. "The Role of Cooperatives in the Development of the Movement Culture of Populism." *Journal of American History* 69, no. 4 (March 1983): 866–85.

Pennefather, R.S. "The Orange Order and the United Farmers of Ontario 1919–1923." *Ontario History* 69, no. 3 (September 1977): 169–84.

Pennington, Doris. *Agnes Macphail: Reformer.* Toronto: Simon and Pierre, 1989.

Perkins, B.B. *Cooperatives in Ontario: Their Development and Current Position.* Guelph: Ontario Agricultural College, 1960.

Phillips, Harry Charles John. "Challenges to the Voting System in Canada, 1874–1974." PhD thesis, University of Western Ontario, London, 1976.

Price, Elizabeth. "The Changing Geography of the Woollen Industry in Lanark, Renfrew and Carleton Counties 1830–1911." MA research paper, University of Toronto, 1979.

Rankin, Pauline. "The Politicization of Ontario Farm Women." In *Beyond the Vote: Canadian Women and Politics*, edited by Linda Kealey and Joan Sangster. Toronto: University of Toronto Press, 1989.

Reeds, L.G. "The Environment." In *Ontario*, edited by Louis Gentilcore. Toronto: University of Toronto Press, 1972.

Richards, John. "The New Populism." *Labour/Le Travail* 23 (Spring 1989): 263–67.

– "Populism: A Qualified Defence." *Studies in Political Economy* 5 (Spring 1981): 5–27.

Roberts, Barbara. "Women's Peace Activism in Canada." In *Beyond the Vote: Canadian Women and Politics*, edited by Linda Kealey and Joan Sangster. Toronto: University of Toronto Press, 1989.

Robertson, James W. *Some Occurrences and Conditions Overseas Which Affect the Production and Marketing of Canadian Agricultural Products: Summary of an Address*. Ottawa: King's Printer, 1920.

Russell, Peter A. "Upper Canada: A Poor Man's Country? Some Statistical Evidence." In *Canadian Papers in Rural History Vol. 3*, edited by Donald H. Akenson. Gananoque: Langdale Press, 1982.

Rutherford, P.F.W. "The People's Press: The Emergence of the New Journalism in Canada, 1869–99." *Canadian Historical Review* 56, no. 2 (June 1975): 169–91.

Samson, Daniel, ed. *Contested Countryside: Rural Workers and Modern Society in Atlantic Canada, 1800–1950*. Fredericton: Acadiensis Press, 1994.

Sandwell, R.W. "Rural Reconstruction: Towards a New Synthesis in Canadian History." *Histoire sociale/Social History* 27, no. 3 (May 1994): 1–32.

Saywell, John T. *'Just Call me Mitch': The Life of Mitchell F. Hepburn*. Toronto: University of Toronto Press, 1991.

Simcoe County Federation of Agriculture. *Report on Cooperatives Activities in Simcoe County.* Barrie: Ontario Department of Agriculture and the Community Life Training Institute, 1944.

Shortt, S.E.D. "Social Change and Political Crisis in Rural Ontario: The Patrons of Industry, 1889–1896." In *Oliver Mowat's Ontario,* edited by Donald Swainson. Toronto: Macmillan of Canada, 1972.

Skey, B.P. "Cooperative Marketing of Agricultural Products in Ontario." PhD thesis, University of Toronto, 1933.

Smart, John David. "The Patrons of Industry in Ontario." MA thesis, Carleton University, Ottawa, 1969.

Staples, Melville H. *The Challenge of Agriculture: The Story of the United Farmers of Ontario.* Toronto: George N. Morang, 1921.

Swierenga, Robert P. "The Malin Thesis of Grassland Accumulation and the New Rural History." In *Canadian Papers in Rural History Vol. V,* edited by Donald H. Akenson. Gananoque: Langdale Press, 1986.

Taylor, Jeffrey M. "The Language of Agrarianism in Manitoba, 1890–1925." *Labour/Le Travail* 23 (Spring 1990): 91–118.

Taylor, Michael. *Anarchy and Co-operation.* Toronto: John Wily and Sons, 1976.

Tennyson, Brian D. "The Ontario General Election of 1919: The Beginnings of Agrarian Revolt." *Journal of Canadian Studies* 4, no. 1 (February 1969): 26–36.

Thomas, R. "The Ideas of W.C. Good, a Christian and Agrarian Reformer: The Formative Years, 1914–1919." MA thesis, University of Ottawa, 1973.

Thompson, John Herd. "Writing About Rural Life and Agriculture." In *Writing About Canada: A Handbook for Modern Canadian History,* edited by John Schultz. Scarborough: Prentice-Hall Canada, 1990.

Thompson, John Herd, with Alan Seager. *Canada 1921–39: Decades of Discord.* Toronto: McClelland and Stewart, 1985.

Trowbridge, R. "Wartime Rural Discontent and the Rise of the United Farmers of Ontario, 1914–1919." MA thesis, University of Waterloo, 1966.

Turner, Larry, with John T. Stewart. *Perth: Tradition and Style in Eastern Ontario.* Toronto: Natural Heritage/Natural History, 1992.

Van Loon, Richard. "Political Participation in Canada: The 1965 Election." *Canadian Journal of Political Science* 3, no. 3 (September 1970): 376–99.

– "The Political Thought of the United Farmers of Ontario." MA thesis, Carleton University, Ottawa, 1965.

Van Loon, Richard, and Michael S. Whittington. *The Canadian Political System: Environment, Structure and Process.* Toronto: McGraw-Hill, 1971.

Voisey, Paul. *Vulcan: The Making of a Prairie Community.* Toronto: University of Toronto Press, 1988.

Walden, Keith. *Becoming Modern in Toronto: The Industrial Exhibition and the Shaping of a Late Victorian Culture.* Toronto: University of Toronto Press, 1997.

Whitaker, Reg. "Images of the State in Canada." In *The Canadian State: Political Economy and Political Power,* edited by Leo Panitch. Toronto: University of Toronto Press, 1977.

– "Introduction." In William Irvine. *The Farmers in Politics.* Toronto: McClelland and Stewart, 1976. First published 1920.

– "Writing about Politics." In *Writing About Canada: A Handbook for Modern Canadian History,* edited by John Schultz. Scarborough: Prentice-Hall Canada, 1990.

White, Randall. *Ontario 1610–1985: A Political and Economic History.* Toronto: Dundurn Press, 1985.

– "The Province of Ontario Savings Office, 1922–1990: A Case Study in the Complexities of Ontario Political Culture." *Ontario History* 87, no. 1 (March 1995): 21–44.

Winson, Anthony. *The Intimate Commodity: Food and the Development of the Agro-Industrial Complex in Canada.* Toronto: Garamond Press, 1993.

Wise, S.F. "The Ontario Political Culture: A Study in Complexities." In *The Government and Politics of Ontario 4th Edition,* edited by Graham White. Scarborough: Nelson Canada, 1990.

– "Upper Canada and the Conservative Tradition." In *Profiles of a Province,* edited by Edith G. Firth. Toronto: Ontario Historical Society, 1967.

Wood, Louis Aubrey. *A History of Farmers' Movements in Canada: The Origins and Development of Agrarian Protest 1872–1924.* Toronto: University of Toronto Press, 1975. First published 1924.

Woodcock, George. *Anarchism: A History of Libertarian Ideas and Movements.* Harmondsworth: Penguin Books, 1962. Reprinted with a postscript, 1975.

– *Anarchism and Anarchists.* Kingston: Quarry Press, 1992.

Wylie, T. Robin. "Direct Democrat: W.C. Good and the Ontario Farm Progressive Challenge, 1895–1929." PhD thesis, Carleton University, Ottawa, 1991.

Young, W.R. "Conscription, Rural Depopulation, and the Farmers of Ontario, 1917–19." *Canadian Historical Review* 53, no. 3, (September 1972), 289–320.

– "The Progressives." In *Readings in Canadian History: Post-Confederation, 3d Edition,* edited by R. Douglas Francis and Donald B. Smith. Toronto: Holt, Rinehart and Winston of Canada, 1990.

Index